Nelson TEC books

CONSTRUCTION TECHNOLOGY

K. Roberts

Lecturer and TEC Course Co-ordinator for B Sector Studies,
South Fields College of Further Education, Leicester

LEVEL 3

Nelson

Thomas Nelson and Sons Ltd
Nelson House Mayfield Road
Walton-on-Thames Surrey KT12 5PL

PO Box 18123 Nairobi Kenya

116-D JTC Factory Building
Lorong 3 Geylang Square Singapore 1438

Thomas Nelson Australia Pty Ltd
19–39 Jeffcott Street West Melbourne Victoria 3003

Nelson Canada Ltd
81 Curlew Drive Don Mills Ontario M3A 2R1

Thomas Nelson (Hong Kong) Ltd
Watson Estate Block A 13 Floor
Watson Road Causeway Bay Hong Kong

Thomas Nelson (Nigeria) Ltd
8 Ilupeju Bypass PMB 21303 Ikeja Lagos

©Keith Roberts 1981
First published 1981

ISBN 0 17 741120 1
NCN 5839 42 0

Design by Kevin Mangan
Filmset by Vantage Photosetting Co. Ltd., Southampton and London
Printed and bound in Hong Kong

Preface

This textbook has been written to cover the TEC standard unit in Construction Technology IIIa. It has been designed to allow the student to learn the terms and concepts which are required by the syllabus in a variety of individual learning methods. The aim has been to provide the means by which learning can be self-paced.

Carefully prepared questions and projects have been incorporated into the text so that the student may check his progress and consolidate what he has learned.

Where appropriate, extracts from the Building Regulations (1976) and Building Research Establishments Digests have been included in the text.

Acknowledgements

The author and publishers wish to acknowledge the help of the following in providing photographs for use in this book.

Pilcon Engineering Ltd (p. 10); Goodenough Pumps Ltd (p. 30 *centre left and right*); Flygt Pumps Ltd (p. 30 *bottom left and right*); J C Bamford Ltd (p. 47, and line drawing p. 48); Priestman Bros. Ltd (p. 48); Benford Ltd (p. 71); Cement and Concrete Association (p. 73); Leicester Mercury (pp. 85, 93); Glow-worm Ltd (p. 195 *bottom left*); Valor Newhome Ltd (p. 195 *bottom right*).

Contents

Building Research Establishment Digests

The following Building Research Establishment Digests have been reproduced.
They are contributed by the courtesy of the Director, BRE.
Crown copyright, reproduced by permission of the Controller, HMSO.

p. 20 BRE Digest 174 *Concrete in sulphate-bearing soils and ground waters*
p. 22 BRE Digest 89 *Sulphate attack on brickwork*
p. 98 BRE Digest 164 *Clay brickwork: 1*
p. 102 BRE Digest 165 *Clay brickwork: 2*
p. 107 BRE Digest 157 *Calcium silicate (sandlime, flintlime) brickwork*
p. 163 BRE Digest 198 *Painting walls Part 2: Failures and remedies*

Building Regulations

Extracts from the Building Regulations 1976 (SI 1676) have been included on the
following pages: 8, 86, 87, 88, 127, 140, 149, 173, 191.
Crown copyright, reproduced by permission of the Controller, HMSO.

1 SOIL INVESTIGATION

THE IMPORTANCE OF SOIL INVESTIGATION

Why do we have to consider ground and soil conditions before developing land for building purposes? The following case study should make this abundantly clear.

A developer received a complaint from the purchaser of a new house, which after two years had developed a serious crack in the front wall. There was a similar crack in the house at the rear, but no other houses on the estate appeared to be affected. As the failure had occurred in only one house, the fault could be bad workmanship and materials.

The builder strongly denied it was due to faulty workmanship and stated all work was carried out to the architect's satisfaction. The builder also pointed out that Building Control had approved the work. The owner was advised to pursue his case in the Law Courts. He asked the Building Research Establishment to investigate the building defect. The bill for this investigation and report was about £500. In their report the Building Research Establishment reported a foundation failure, as indicated below:

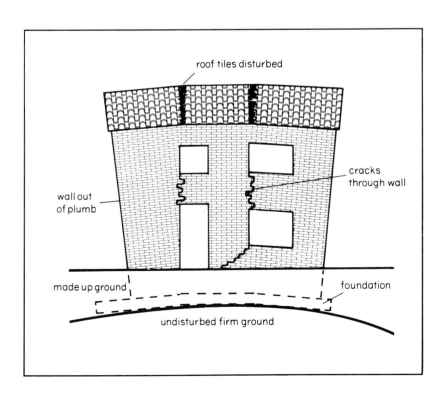

The house owner on advice from his lawyer proceeded against the builder. The judge in summing up asked the question, 'Would a competent person have foreseen this potential damage?' Judgement was awarded against the builder.

How could the builder/architect have obtained that now expensive information?

Having learnt from other people's mistakes you are proposing to develop a plot of land for an office block. The plot is very attractively situated with four mature trees. The architect has cleverly landscaped the building to get the best potential from the existing land environment. He must also be clever enough to make sure the building doesn't crack or collapse.

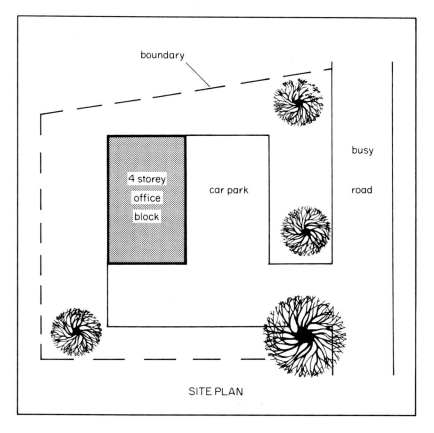

SITE PLAN

The foundations must be adequate for a start. His calculations must be right. The **building load** must be worked out, and the **safe bearing capacity** of the soil must be found out. Then the architect can work out what sort of area the foundations should cover, using the following simple formula,

$$\text{foundation area (or foundation spread)} = \frac{\text{building load (kN)}}{\text{safe bearing capacity of soil (kN/m)}}$$

Minimum width of strip foundations

Type of subsoil (1)	Condition of subsoil (2)	Field test applicable (3)	Minimum width in millimetres for total load in kilonewtons per lineal metre of loadbearing walling of not more than–					
			20 kN/m (4)	30 kN/m (5)	40 kN/m (6)	50 kN/m (7)	60 kN/m (8)	70 kN/m (9)
I Rock	Not inferior to sandstone, limestone or firm chalk	Requires at least a pneumatic or other mechanically operated pick for excavation	In each case equal to the width of wall					
II Gravel Sand	Compact Compact	Requires pick for excavation. Wooden peg 50 mm square in cross-section hard to drive beyond 150 mm	250	300	400	500	600	650
III Clay Sandy clay	Stiff Stiff	Cannot be moulded with the fingers and requires a pick or pneumatic or other mechanically operated spade for its removal	250	300	400	500	600	650
IV Clay Sandy clay	Firm Firm	Can be moulded by substantial pressure with the fingers and can be excavated with graft or spade	300	350	450	600	750	850
V Sand Silty sand Clayey sand	Loose Loose Loose	Can be excavated with a spade. Wooden peg 50 mm square in cross-section can be easily driven	400	600	Note: Foundations do not fall within the provisions of regulation D7 if the total load exceeds 30 kN/m			
VI Silt Clay Sandy clay Silty clay	Soft Soft Soft Soft	Fairly easily moulded in the fingers and readily excavated	450	650	Note: In relation to types VI and VII, foundations do not fall within the provisions of regulation D7 if the total load exceeds 30 kN/m			
VII Silt Clay Sandy clay Silty clay	Very soft Very soft Very soft Very soft	Natural sample in winter conditions exudes between fingers when squeezed in fist	600	850				

D7   Deemed-to-satisfy provisions for strip foundations

If the foundations of a building are constructed as strip foundations of plain concrete situated centrally under the walls, the requirements of regulations D3(a) shall be deemed to be satisfied if–

(a) there is no made ground or wide variation in the type of subsoil within the loaded area and no weaker type of soil exists below the soil on which the foundations rest within such a depth as may impair the stability of the structure;

(b) the width of the foundations is not less than the width specified in the Table to this regulation in accordance with the related particulars specified in this Table;

(c) the concrete is composed of cement and fine and coarse aggregate conforming to BS 882: Part 2: 1973 in the proportion of 50 kg of cement to not more than 0·1 m³ of fine aggregate and 0·2 m³ of coarse aggregate;

(d) the thickness of the concrete is not less than its projection from the base of the wall or footing and is in no case less than 150 mm;

(e) where the foundations are laid at more than one level, at each change of level the higher foundations extend over and unite with the lower foundations for a distance of not less than the thickness of the foundations and in no case less than 300 mm; and

(f) where there is a pier, buttress or chimney forming part of a wall, the foundations project beyond the pier, buttress or chimney on all sides to at least the same extent as they project beyond the wall.

SITE INVESTIGATION

The building load can be calculated in the office, but what needs to be found out on the site? The bearing capacity of the soil and its depth.

For house building loads the Building Regulations Table D7 can be used. A trial pit will provide a **field classification** and an approximate **bearing classification**. A field test will determine whether clay is soft or firm.

Read through the Building Regulations opposite and using Table D7 look up the classification of soil from these field tests:

    a) the clay was moulded by substantial pressure with fingers

    b) the clay was easily moulded by fingers.

One only has a limited number of fingers to break.

Surely there must be a better and safer way of obtaining soil bearing pressure. The field test suggested above is satisfactory for houses, but what of buildings like office blocks or multi-storey flats? The bigger the proposed building the more accurate the assessment of the soil bearing pressure should be.

The type of building (such as loadbearing brick or reinforced concrete frame), is another factor determining the extent of the soil investigation. This is called 'nature of work'.

Review

Let us review our progress so far.

1   What will you need to know before you can work out the foundation area?

2   What does 'safe bearing capacity of the soil' mean?

3   What two factors affect the scope of soil investigation?

WHAT TO LOOK FOR

The soil investigation is necessary to obtain essential information such as

a) type of soil (e.g. clay or sand)
b) depth of each different type of soil (called **soil strata**)
c) the bearing capacity of the soil strata
d) the moisture content of the different soil strata.

Finding this information can be expensive so we use **soil samples**, which reduce the cost of soil investigation and give a good guide to soil conditions.

Consider the proposed plot for the office. *Where on the site plan would you locate the soil samples?*

We can assume the continuity of soil strata between the trial pits, and for this reason the samples should be located at each corner of the building.

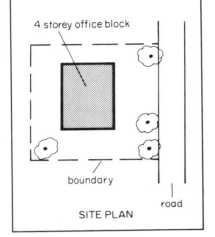

SITE PLAN

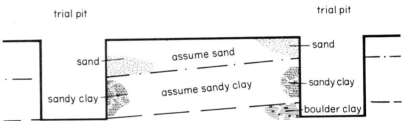

Can we reduce the amount of excavation during the soil investigation? Often the depth required to obtain soil samples far exceeds four metres and makes the digging of trial pits very expensive. What is an alternative to trial pits? **Auger** or **shell borings** are acceptable alternatives. Instead of excavating a pit, a column of earth is removed and a record of the depth and type of soil is entered into a bore hole log.

The equipment used comprises a winch, shear legs and a split spoon sampler or basket if in gravels and sands.

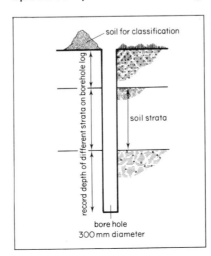

The rest of our soil investigation will be done in a laboratory analysis of the soil samples. Not all the soil obtained needs to be analysed, only a representative sample. Two separate samples are sent to the laboratory for analysis.

1 **Disturbed** sample
2 **Undisturbed** sample

An undisturbed sample has the same moisture content as in the ground. This is important because the moisture content of the soil can affect the bearing pressure of the soil. The depth of the undisturbed sample needs to be recorded on the bore hole log.

From the bore hole log shown two samples of soil are obtained, one a disturbed sample at 1·00 metre depth in sand and the other an undisturbed sample at 2·50 metres depth in the sandy clay. *What will happen to the moisture content of the sample on exposure to the air?*

The sample needs to be labelled and sealed in a container to prevent any loss of moisture. Only then will it be an undisturbed sample. The moisture content and the structure of the soil will remain unaltered.

Both samples, the disturbed and undisturbed, are sent to the laboratory for analysis. The soil laboratory will

1 classify the soil type
2 assess the bearing pressure
3 assess the moisture content.

For the latter two tests, the bearing pressure and moisture content, the laboratory will use the undisturbed sample. Having obtained a satisfactory assessment of soil conditions the architect can proceed with detailed design confidently.

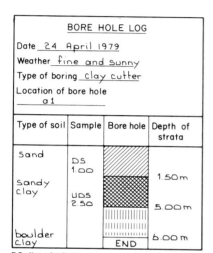

BORE HOLE LOG

Date 24 April 1979
Weather fine and sunny
Type of boring clay cutter
Location of bore hole
        a 1

Type of soil	Sample	Bore hole	Depth of strata
Sand	DS 1.00		
Sandy clay	UDS 2.50		1.50 m
			5.00 m
boulder clay		END	6.00 m

DS disturbed sample
UDS undisturbed sample

Additional reading

BRE Digest 64 (*Soils and Foundations : 2*)

Review

1 Where approximately would the bore holes be positioned for the proposed four storey office block?
2 What four pieces of information do we need to obtain from the bore hole samples?
3 What is the record called for keeping this information?
4 What information is readily available from the bore hole?
5 Explain the difference between, and the laboratory use of, a disturbed and an undisturbed sample.

2 RETAINING WALLS

As the name 'retaining wall' implies, its primary function is to contain soil. 'Soils' vary from those, like rock, which can support themselves to non-cohesive soils, such as running sand, which behave like liquids and have to be supported.

Such soils when excavated and left unsupported will slip from the vertical face to a natural angle or slope. This slope is called **angle of friction**. Different soils will have differing angles of friction.

Angle of friction for different soils

Angle of friction	Soil type
45°	Vegetable soil Moist rubble Moist clay
30°	Dry vegetable soil Dry sand Dry clay

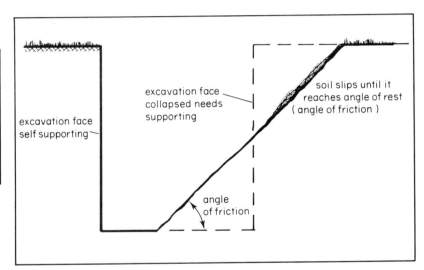

If the excavation or soil face needed retaining what pressures would the retaining wall have to resist?

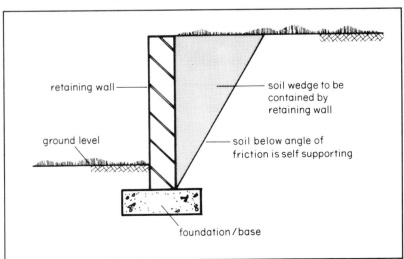

FORCES TO BE RESISTED

First, the retaining wall will have to resist the pressure exerted by the earth wedge. Second, ground water will put additional pressure on the wall and reduce its stability. This is because water behind the retaining structure will increase lateral pressure and may reduce the bearing capacity of the ground under its base. The soil friction which helps stop sliding at the foundation of the wall may be reduced, and the durability of the material in the retaining wall will be affected.

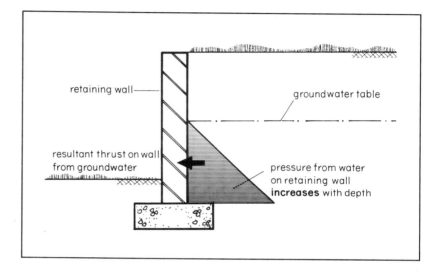

Third, imposed loads will increase the pressure already exerted by the earth wedge.

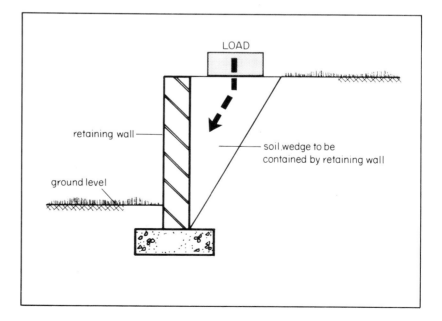

Review

Answer the following questions.

1 What is the primary function of an earth retaining wall?
2 Explain the angle of friction as it affects the pressure on the retaining wall.
3 What is the angle of friction or repose for the following soils? *a*) Dry clay *b*) Dry sand *c*) Moist clay
4 List the loads or pressures the retaining wall has to resist.

Now check your answers before going on.

MASS RETAINING WALLS – MAKING THEM STABLE

Retaining walls are usually **free standing** unless acting as part of a large structure (e.g. basement-wall).

They can be classified as

1 mass or gravity walls
2 flexible cantilever walls
3 counterfort walls.

In this section we are only concerned with the mass retaining walls. Mass retaining walls resist overturning and bending stresses largely because of their mass and are generally made out of brick, stone or plain concrete up to 2 m high. Let us now consider how retaining walls can fail.

The wall tends to move or slip due to pressure of the earth. If the wall is to be stable it must exert an equal or greater opposite force. The resisting force is the mass of the wall and, to a limited extent, the soil at the base.

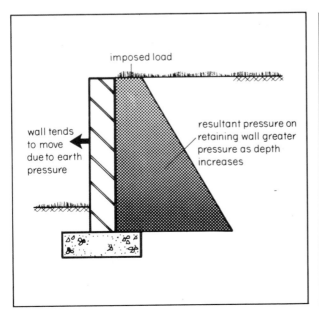

imposed load

wall tends to move due to earth pressure

resultant pressure on retaining wall greater pressure as depth increases

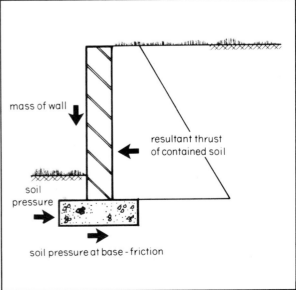

mass of wall

resultant thrust of contained soil

soil pressure

soil pressure at base - friction

What happens if there is insufficient resistance to the earth pressure? The wall fails by sliding.

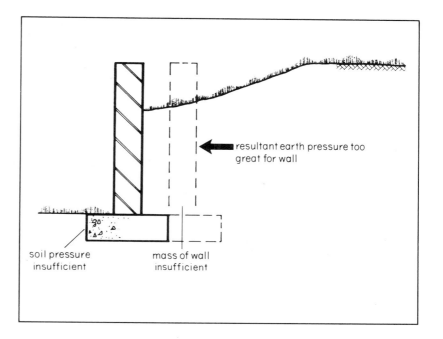

This can be prevented by the passive resistance of the soils at the front of base and increasing this passive resistance by providing a toe projection underneath the base.

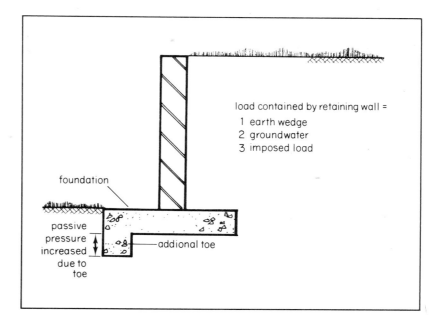

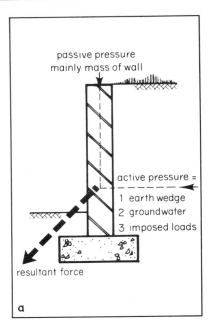

passive pressure
mainly mass of wall

active pressure =
1 earth wedge
2 groundwater
3 imposed loads

resultant force

a

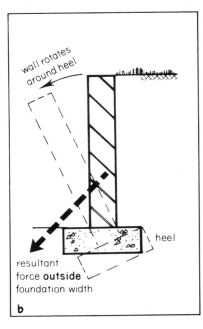

wall rotates
around heel

heel

resultant
force **outside**
foundation width

b

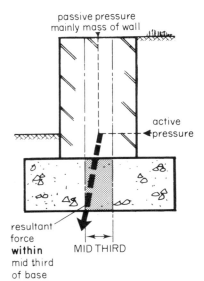

passive pressure
mainly mass of wall

active
pressure

resultant
force
within
mid third
of base

MID THIRD

The retaining wall may also fail due to rotation. The forces acting on the retaining wall can be illustrated as a vector diagram. (Fig. **a**)

The resultant force of these two vectors is important for the stability of the wall. If the resultant force is outside the base of the wall, the wall will fail due to rotation. (Fig. **b**)

To avoid this the **mid third rule** is used in calculations. This states that the resultant force must be within mid third of base.

The other effect could be to overstress the safe bearing capacity of soil under the wall base.

Review

1   What three forces allow the retaining wall to resist slipping?
2   What rule is applied to prevent the retaining wall rotating?

Check your answers before going on.

RETAINING WALL MATERIALS

Now we know some of the terms used in retaining walls, we can look at the materials to be used. Obviously the materials have to be carefully selected if they are to do their job satisfactorily.

There can be problems with the materials used in retaining walls, which BRE Digest 174 (*Concrete in sulphate-bearing soils and groundwaters*) and BRE Digest 89 (*Sulphate attack on brickwork*) highlight.

Building Research Establishment Digest 174

Concrete in sulphate-bearing soils and groundwaters

Factors influencing attack

Aqueous solutions of sulphates attack the set cement in concrete, the chemical reactions occurring depending upon the kind of sulphate present and the nature of the cement. The rate of attack depends greatly on the permeability of the concrete. The factors influencing attack are:

(1) the amount and nature of the sulphate present;
(2) the level of the water table and its seasonal variation;
(3) the form of construction;
(4) the type and quality of the concrete.

If sulphate conditions cannot be prevented from reaching the structure, the only defence against attack lies in the control of the fourth factor; it cannot be too strongly emphasised that fully compacted concrete of low permeability must be used.

Occurrence of sulphates Sulphates occur mainly in strata of London Clay, Lower Lias, Oxford Clay, Kimmeridge Clay and Keuper Marl. The most abundant salts are:

> calcium sulphate (gypsum or selenite)
> magnesium sulphate (Epsom salt)
> sodium sulphate (Glauber's salt).

Sulphuric acid and sulphates in acid solution occur much less often, for example, near colliery tips, in marshy country and where pyrite in the soil is being slowly oxidized.

Sulphates are sometimes present in materials such as colliery shale used as fill beneath solid concrete ground floors. Moisture from the ground carries the salts to the underside of the concrete, which they attack, leading to expansion and cracking of the floor and the surrounding structure.

Brick rubble, particularly with adhering plaster, ashes and some industrial wastes, are potential sources of damage by sulphate attack.

The solubilities of the salts mentioned – expressed as grams of sulphur trioxide (SO_3) per litre of solution – vary considerably. That of calcium sulphate is only 1.2 g (SO_3) per litre, compared with about 150 and 200 times this value for magnesium and sodium sulphates respectively. The SO_3 content of groundwaters therefore gives an indication that one or both of the latter sulphates may be present.

Movement of groundwater Sulphates can only continue to reach concrete by movement of their solutions in water. Thus concrete which is wholly and permanently above the water table is unlikely to be attacked, even though capillarity in the subsoil may cause some migration of salts to zones above the water table. Below the table, however, movement of water may replenish the sulphates removed by reaction with concrete, thereby maintaining the attack. Movements may be vertical or horizontal, depending on seasonal variations in rainfall and on the geology of the site and its environs. A low sulphate content in the soil does not eliminate the possibility of sulphate attack, since groundwater may flow from adjacent areas, particularly if the soil has been disturbed, e.g. by laying pipes.

Form of construction Given the same concrete quality and corrosive conditions, massive forms of construction will, as expected, deteriorate less quickly than thin sections. Special considerations apply to piles, since if the surface skin of concrete is attacked there could be changes in the friction on the pile shaft. Precast piles have advantages over those cast *in situ*; concrete quality – in particular compaction – is more easily controlled, since placing is less difficult (*see* Digest 95).

In basement structures, retaining walls, or culverts, for example, the concrete may have to resist strong lateral water pressure on one side only. This favours both water penetration and concentration of salts on the other side if evaporation of water from that side can occur. Similarly, partly buried concrete may draw up sulphates in solution from beneath by a combination of capillarity and evaporation at the exposed surfaces.

At first sight precast concrete pipes may appear rather more vulnerable to attack than other forms of concrete; they are thin in section and may be subjected to water pressure from one side only. Against this, however, is the fact that the concrete quality is usually high, to satisfy BS 556, and, providing the appropriate cement is used, this is the best safeguard against attack. (The internal corrosion of concrete pipes by sulphate salts in the effluent is discussed in Note No. 6: 1959 by the Water Pollution Research Laboratory).

Type and quality of concrete Concrete of low permeability is essential to resist sulphate attack; hence the concrete must be fully compacted. The mix must be so designed that it has sufficient workability to allow full compaction and yet has a free water/cement ratio no greater than that specified for the particular soil conditions. 'Free' water is the total weight of water in the concrete less that absorbed by the aggregate. (A method of measuring water absorption of aggregates is given in BS 812, Section 4).

Where sulphates are present, the mix should be designed to have a cement content not less than, and a free water/cement ratio not greater than, the values specified in Table 1 as appropriate to the severity of the conditions and the type of cement. For a particular situation, the cement content may have to be greater than the minimum specified in the table, so as to achieve the required workability within the limits of the maximum free water/cement ratio. This is particularly so for *in situ* concrete piles where the overriding consideration is to ensure complete compaction and therefore structural integrity of the pile; in this case, the mix must be designed to produce a highly workable but cohesive concrete.

Admixtures for concrete such as air-entraining or water-reducing agents may give some limited improvement in sulphate resistance. Admixtures containing workability aids improve compaction and allow the use of lower water/cement ratios, but those containing calcium chloride are not recommended.

All precast units and any type of concrete using supersulphated cement require particular care during the initial curing period, so as to ensure minimum permeability and a hard surface.

Types of cement

In concrete prepared from ordinary or rapid-hardening **Portland cement** (BS 12) one reaction which occurs in sulphate solutions is with the hydrates formed from any tricalcium aluminate in the cement. Another involves the calcium hydroxide released during hydration. These reactions lead to a considerable expansion in the cement, finally resulting in the breakdown of the concrete. A considerable resistance to sulphate attack is obtained when the tricalcium aluminate content of the cement is kept low. Portland cements complying with BS 12 usually contain a significant proportion of tricalcium aluminate, but in **sulphate-resisting Portland cement** complying with BS 4027 the amount is limited to 3·5 per cent.

The resistance of concrete made with **supersulphated cement** (BS 4248) to attack by sulphates over the range of concentrations commonly encountered in groundwaters is similar to that of sulphate-resisting Portland cement. This cement is not currently produced in the UK but it has been found that good quality concrete made with it has an acceptable life in mildly acid conditions. Laboratory tests indicate that when immersed in strong magnesium sulphate solution (up to 3·5 per cent SO_3) concrete samples prepared from supersulphated cement lose strength more quickly than those prepared from sulphate-resisting Portland cement. In strong sodium sulphate solution, the current evidence is divided but, on balance, the relative resistances appear to be reversed.

The sulphate resistance of cements based on mixtures of granulated blastfurnace slag and Portland cement clinker increases with the proportion of slag. In addition to factory-produced cements, ground granulated blastfurnace slag is available for introduction at the mixer. BS 146 for **Portland-blastfurnace cement** does not set a minimum level for slag content and this cement is therefore classified with ordinary Portland cement. BS 4246 for **low heat Portland-blastfurnace cement** specifies a minimum slag content of 50 per cent which is probably less than is required for any substantial improvement in sulphate resistance over ordinary Portland cement but until the result of long-term tests are available there is little UK data on which to base precise recommendations.

The inclusion of pozzolanas improves the sulphate resistance of ordinary Portland cement concretes. Selected and classified pulverised-fuel ashes with good pozzolanic properties are available in the UK but here again the results of long-term tests are awaited.

Classification of sites and recommendations for concrete

For classification purposes sites have been divided into five categories of increasing severity, based on the sulphate contents of the soil and groundwater. The recommended type of cement and the minimum cement content for each of these classes are given in Table 1. It must be remembered that the divisions between the classes are somewhat arbitrarily drawn and that the recommendations are judgments based on present knowledge.

Sampling and analysis of soils and groundwaters

All reliable information about the site should be examined in assessing the need for extensive analysis of the soil. The information should include an estimate of the sulphate content of any groundwater samples obtained during site investigation. The main danger in classifying sites on the basis of groundwater analysis alone lies in the difficulty in obtaining samples that are not diluted with surface water.

Suitable soil samples may be obtained from the test boreholes made for engineering purposes. They should be taken at every 1–2 m and wherever an obvious change in stratum occurs. Economic considerations will govern the number of soil samples analysed. When the cost of analysis is low compared with the preparation and bulking of the sample it is prefer-

Table 1   Requirements for concrete exposed to sulphate attack

	Concentration of sulphates expressed as SO_3				Requirements for dense, fully compacted concrete made with aggregates meeting the requirements of BS 882 or BS 1047			
	In soil		In ground-water	Type of cement	Minimum cement content Nominal maximum size of aggregate			Maximum free water/cement ratio
Class	Total SO_3	SO_3 in 2:1 water:soil extract			40 mm	20 mm	10 mm	
1	% less than 0·2	g/litre —	parts per 100,000 less than 30	Ordinary Portland or Portland-blastfurnace	kg/m³ 240	kg/m³ 280	kg/m³ 330	0·55
2	0·2–0·5	—	30–120	Ordinary Portland or Portland-blastfurnace	290	330	380	0·50
				Sulphate-resisting Portland	240	280	330	0·55
				Supersulphated	270	310	360	0·50
3	0·5–1·0	1·9–3·1	120–250	Sulphate-resisting Portland or supersulphated	290	330	380	0·50
4	1·0–2·0	3·1–5·6	250–500	Sulphate-resisting Portland or supersulphated	330	370	420	0·45
5	over 2	over 5·6	over 500	As for Class 4, but with the addition of adequate protective coatings of inert material such as asphalt or bituminous emulsions reinforced with fibreglass membranes				

Notes

This table applies only to concrete made with aggregates complying with the requirements of BS 882 or BS 1047 placed in near neutral groundwaters of pH 6 to pH 9, containing naturally occurring sulphates but not contaminants such as ammonium salts. Concrete prepared from ordinary Portland cement would not be recommended in acidic conditions (pH 6 or less); sulphate-resisting Portland cement is slightly more acid-resistant but no experience of large-scale use in these conditions is currently available. Supersulphated cement has given an acceptable life, provided that the concrete is dense and prepared with a free water/cement ratio of 0·40 or less, in mineral acids down to pH 3·5.

The cement contents given in Class 2 are the minima recommended by the manufacturers. For SO_3 contents near the upper limit of Class 2, cement contents above these minima are advised.

For severe conditions, eg thin sections, sections under hydrostatic pressure on one side only and sections partly immersed, consideration should be given to a further reduction of water/cement ratio and, if necessary, an increase in cement content to ensure the degree of workability needed for full compaction and thus minimum permeability.

able to analyse the samples separately. If, to save costs, samples are combined, this should only be done for samples from the same depth in different boreholes when the type of stratum is the same.

Sulphate contents of the groundwater can be determined by any suitable analytical method. Concentrations in the fourth column of Table 1 are customarily expressed as parts SO_3, per 100 000 parts of water, so that analytical figures expressed in grams of SO_3 per litre must be multiplied by 100. The total sulphate content of the soil may be obtained by extraction with hot dilute hydrochloric acid, and is expressed as the percentage of SO_3 per dry weight of soil. If the sulphate present in the soil is predominantly the calcium salt, its low solubility may result in the total sulphate content giving too severe a classification for the site. In cases where the total sulphate exceeds 0·5 per cent it is suggested that the water-soluble sulphate should be determined. The soluble sulphate salts can usually be extracted from the soil using twice the weight of water and expressed as grams of SO_3 per litre of water extract. Classification on this basis is given in column 3 of Table 1.

The detailed analytical procedures for these determinations are published in a BRE Current Paper*, together with the methods for the quantitative determination of the metallic ions present.

Interpretation of analytical results   The results of all the individual chemical analyses should be available to the engineer to assist him in deciding on the precautions necessary. If classification is based solely on the analysis of groundwaters, it should correspond to the highest sulphate concentration recorded. If classification is based on the analysis of only a small number of soil samples and the results vary widely, it may be worth while to take further samples for analysis. When a larger number of results are available it is suggested that the site classification should be based on the mean of the highest 20 per cent of the results. When the soil samples have been combined before analysis the selection should be more stringent and the mean of the highest 10 per cent of results is suggested.

British Standards referred to in the text

BS 12   *Portland cement (ordinary and rapid-hardening)*
         Part 2: 1971 Metric units

BS 146: 1958 *Portland-blastfurnace cement*

BS 556   *Concrete cylindrical pipes and fittings including manholes, inspection chambers and street gullies*
         Part 2: 1972 Metric units

BS 812   *Methods of sampling and testing of mineral aggregates, sands and fillers*

BS 4027 *Sulphate-resisting Portland cement*
         Part 2: 1972 Metric units

BS 4246 *Low heat Portland-blastfurnace cement*
         Part 2: 1974 Metric units

BS 4248: 1974 *Supersulphated cement*

*CP3/68 *Analysis of sulphate-bearing soils in which concrete is to be placed.* S. R. Bowden.

Review

Read carefully BRE Digest 174 (*Concrete in sulphate-bearing soils and groundwaters*) and answer the following questions.

1 The rate of sulphate attack depends greatly on the permeability of something. What is it?
2 What are the four factors influencing attack?
3 How do sulphates continue to reach concrete?
4 Is concrete which is wholly or permanently above the water table likely to be attacked?
5 Does a low sulphate content in the soil eliminate the possibility of attack?
6 Should concrete have a high or low permeability in order to resist sulphate attack?
7 What is the most suitable mix of concrete?
8 Admixtures containing workability aids improve compaction and allow the use of low water/cement ratios. Should admixtures containing calcium chloride be used where there is a possibility of sulphate attack?
9 In concrete prepared from Ordinary Portland Cement one reaction which occurs in sulphate solutions is with the hydrates formed from any tricalcium aluminate in the cement. What do these reactions lead to?
10 To obtain resistance to sulphate attack should the tricalcium aluminate content be kept low or high?
11 List four types of cement which can be used in sulphate bearing soils and groundwater.
12 Copy out the table below and fill it in. Recommend the type of cement for these concentrates of sulphate expressed as SO_3 in groundwater.

SO_3 in groundwater parts per 100,000	Type of cement	Minimum cement content 20 mm aggregate	Maximum free water/cement ratio
Less than 30			
30–120			
120–250			
250–500			

Building Research Establishment Digest 89

Sulphate attack on brickwork

The tricalcium aluminate constituent in ordinary Portland cements can react with sulphates in solution to form a compound called calcium sulphoaluminate or 'Ettringite', the reaction being accompanied by expansion. When it occurs in brickwork mortars, the effect is first an overall expansion of the brickwork, which can be followed later in more extreme cases by progressive disintegration of the mortar joints.

Except for the special case of earth-retaining walls in brickwork, where the attacking sulphates could be derived from the groundwater, the source of sulphates is usually the soluble salts present to varying extents in most types of clay brick. The sulphates can only be transferred from the bricks to the mortar joints by percolating water – usually incident rainwater – so that sulphate attack is normally confined to external brickwork. Renderings based on Portland cement may also be affected by sulphates, though their failure when used on high-sulphate bricks is more frequently due to attack and expansion of the backing brickwork. This expansion is usually a secondary effect following the entry of water into the brickwork through defects or shrinkage cracks in the rendering.

Facing brickwork

Initially, the only evidence of sulphate attack may be horizontal cracking on the inner face of the wall, due to this being put in tension by the expansion of the outer leaf. At this stage, the mortar of the outer leaf shows little sign of damage, and may even have hardened slightly. In normal cavity work with metal ties the horizontal cracking is usually concentrated near the roof but in solid brickwork with brick ties cracking may also occur lower down (Fig 1).

Sometimes differential movement can occur between the faces of brickwork; the consequent bowing of walls is a means of relieving the compression. In long stretches of brickwork some oversailing of the damp-proof course usually occurs since in practice sulphate expansion is usually less below the dpc, possibly due to restraint from the foundations (Fig 2). (This oversailing can usually be distinguished from a similar effect due to moisture expansion of the bricks by considering the time factor, discussed later.) As sulphate attack proceeds, the mortar joints take on a whitish appearance, and a narrow crack may appear in the middle of the joints (Fig 3). Later still, the surface of the joint spalls off, and the mortar is reduced to the strength and consistency of a very weak lime mortar. Spalling of facing bricks occasionally accompanies the advanced stages of attack, where comparatively friable bricks have been involved; this is usually a secondary effect due to transfer of excessive load on to the outer leaf of brickwork, which in turn is concentrated on the outer faces of the bricks where the mortar joints have expanded most (Fig 4).

Chimney stacks

In the past many cases have been reported of sulphate attack on unlined chimney stacks serving slow-burning appliances. The provision of liners considerably reduces the risk of attack which is caused mainly by condensation from flue gases (*see* Digest 60).

Fig 1

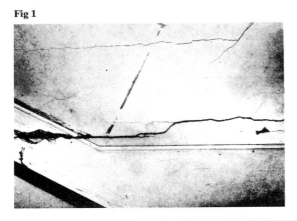

Fig 2

Fig 3

Fig 5

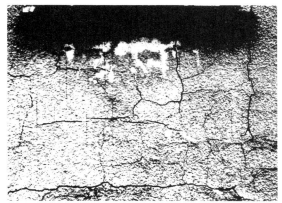

Fig 4

Rendered brickwork

Except in positions where very high rainfall is combined with little opportunity for drying out, properly designed and executed renderings (to BS CP 221: 1960) will prevent the penetration of rain into the underlying brickwork. However, if rain can penetrate the rendering through cracks or poor detailing, it may become trapped and saturate the brickwork for long periods. Sulphate attack may result causing expansion of the brickwork (Fig 5), and extensive areas of rendering may become detached. Sulphate crystals may be found adhering to the back faces of pieces of detached rendering. Usually, the rendering itself remains comparatively unaffected chemically; it is the expansion of the backing brickwork which has caused it to become detached. The fine map cracking due to drying shrinkage of a rendering is easily distinguishable from the horizontal and vertical cracking associated with expansion of mortar joints from sulphate attack.

The time factor

Only in exceptional circumstances is sulphate expansion seriously in evidence in under two years. Hence the oversailing at damp-proof course level due to sulphate attack can usually be distinguished from that caused by moisture expansion of bricks, since the latter is only likely to occur in the first few months after a building is erected.

Amount of expansion

A vertical expansion of up to 0·2 per cent is commonly found with sulphation of the mortar in facing brickwork, although in rare circumstances with rendered work, as much as 2 per cent has been seen.

Conditions necessary for sulphate attack

For sulphate attack to occur, three materials must be present simultaneously: tricalcium aluminate, soluble sulphates and water.

Tricalcium aluminate – This is present in ordinary Portland cement in amounts from under 8 per cent to over 13 per cent. Since, in general, the greater the amount of tricalcium aluminate in a cement the lower will be its resistance to sulphate attack, the susceptibility of ordinary Portland cement is somewhat variable; some varieties have quite high resistance. There is, however, no easy way whereby the user can estimate susceptibility (the brand-name is no guide), so that it must be assumed that all ordinary Portland cements are capable of being attacked. Sulphate-resisting Portland cements, on the other hand, have a very low tricalcium aluminate content and give a positive guarantee of resistance. Supersulphated and high alumina cements also have high resistance to sulphates.

It cannot be assumed that the greater the amount of reactive cement in a mortar mix the more liable it will be to attack; in practice a rich mix such as $1:\frac{1}{2}:4\frac{1}{2}$ is much more resistant than the weaker, more porous $1:1:6$ mix. The reason for this is not fully understood, though the porosity probably has an important effect.

Soluble sulphates. These are present in almost all fired clay bricks, although the amount may vary widely between different types and even between individual bricks of the same type. BS 3921 limits the amount which can be present in bricks of 'special quality' but does not specify maximum permissible amounts of sulphate for bricks of 'ordinary quality'. In all but the most abnormal conditions of design and exposure 'special quality' bricks should never promote sulphate attack. Most bricks, however, cannot meet the sulphate limitations of the 'special quality' classification, and unless the brick manufacturer can produce evidence, in the form of control charts of sulphate content, that his bricks consistently satisfy the British Standard limits in this respect, it must be assumed that they may contribute to sulphate attack. (BS 3921 is discussed in greater detail in Digests 164 and 165.)

Water. This is always present in brickwork during construction, but although often taking considerable time to dry out it appears insufficient by itself to cause sulphate attack. It is good practice to avoid saturation of brickwork during construction in order to limit troubles due to efflorescence. Repeated wetting and drying over a period of years is closely correlated with sulphate attack, the severity of which depends on the exposure to wet weather. Parapets and free-standing walls, which tend to get wet more often than normally protected walls between eaves and foundations, are more likely to be affected. Exceptionally, in regions exposed to driving rain, all brickwork is subject to considerable wetting; some guidance to the worst areas of driving rain can be found in Digest 127.

The most vulnerable form of brickwork is the earth-retaining wall, where water is sometimes allowed to pass continuously through the brickwork from the earth fill. Evidently the amount of water movement through brickwork is an important factor determining the susceptibility to attack. It should be noted that garden walls commonly retain some earth and so are frequently attacked.

Preventing sulphate attack in new work

Ideally, low sulphate content bricks that satisfy the provisions of BS 3921 for soluble salts should be used, but this is a perfection that may either not be justified or possible. In general terms extra precautions should be taken either to increase the resistance of the mortars and renderings to sulphate attack or to limit the extent to which the brickwork becomes and remains wet, eg by improved design, as discussed later; in cases of exceptional risk (for example, in areas of excessive driving rain) both methods should be applied.

Sulphate-resisting mortars and renderings

The sulphate resistance of mortars can be increased in two ways: (1) by specifying the richer mixes, for example: $1:0-\frac{1}{4}:3$ or $1:\frac{1}{2}:4-4\frac{1}{2}$, or (better still) $1:5-6$ with plasticiser in place of lime, and (2) by using sulphate-resisting Portland, supersulphated or high alumina* cements.

The sulphate resistance of renderings can be improved by the use of sulphate-resisting Portland cement. It is of greater importance, however, to avoid the penetration of moisture to the backing brickwork and to allow any that does penetrate to evaporate. Renderings based on strong mixes, particularly if given a trowelled finish, are prone to shrinkage cracking and reduce the drying rate of the brickwork. Aerated renderings may allow rain penetration into the brickwork, especially if used on absorptive bricks.

Improved design

In most areas, undue wetting of the structure can be avoided by attention to design details. For facing brickwork, accepted (though by no means universally employed) measures are:

1 To provide generous overhang at eaves and verges;
2 To provide flashings and damp-proof courses at cills and elsewhere so as to prevent ingress of water from above;
3 To avoid parapets and free-standing walls unless special precautions are taken. These are:
   (a) To use low-sulphate bricks;
   (b) To design copings with generous overhang and adequate drip, with damp-proof courses under;
   (c) To provide damp-proof courses at bases of free-standing walls, i.e. above expected soil levels, and at roof level in parapets;
   (d) To provide expansion joints not more than 12 m apart;
   (e) To specify mortar mixes in accordance with the previous section.

---

*High alumina cements should not be used with lime additions, though ground limestone may be used.

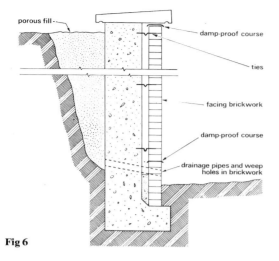

porous fill

damp-proof course

ties

facing brickwork

damp-proof course

drainage pipes and weep holes in brickwork

Fig 6

Earth-retaining walls should be built only in fired clay brickwork if bricks wholly conforming to the 'special quality' of BS 3921 are used in conjunction with sulphate-resisting mortar mixes (eg 1 : $\frac{1}{2}$: $4\frac{1}{2}$ or stronger, and preferably containing a sulphate-resisting cement). If bricks containing appreciable amounts of sulphates must be used, as for example where it is necessary to match the remainder of a building scheme, the retaining wall can be built of *in situ* concrete, without a batter, with the facing base layer separated from it by a cavity (Fig 6). Adequate copings and expansion joints should be provided as for parapet walls, and weep holes should be incorporated at the base of the cavity, taking care to ensure that any drainage also provided for the retained earth does not vent into the cavity.

Remedying existing sulphate attack

Where brickwork has already been affected by sulphates the remedies to be applied depend on the extent of the damage, but all are based on the need to allow the work to dry out and thereafter to exclude moisture as far as possible. In all cases obviously poor design features must be remedied by taking the precautions outlined for new work.

Where expansion affects only particularly vulnerable details such as parapets it may be sufficient merely to correct the detailing and to cut one or two expansion joints in appropriate places.

Where attack is slightly more severe, but has taken some years to develop, and shows only as an expansion of the exterior walling without visible damage to the mortar, an experimental application of surface waterproofer may be tried, applying this to the brickwork when it is reasonably dry. Similar applications have been successful in minimising further expansion, though there is a slight risk of soft-fired bricks spalling because of crystallisation of sulphates behind the face. The waterproofer will have to be renewed at intervals of a few years in accordance with the manufacturer's recommendations (*see* Digest 125).

Where the sulphate attack is advanced and, in addition to brickwork expansion, the mortar is already showing severe damage, some form of cladding should be applied. Weather-boarding or tile hanging are suitable. Similar action should be taken where renderings have failed extensively; it is not sufficient merely to replace the rendering, which is likely to fail again due to the continued expansion of the backing brickwork. Wherever it is necessary to rebuild features completely, every effort should be made to use materials which are more suitable to the conditions.

Review

Read carefully the BRE Digest 89 (*Sulphate attack on brickwork*) and answer the following questions.

1 What is Ettringite?
2 How is Ettringite formed?
3 When sulphate attack occurs in brickwork mortars the first effect is an overall expansion of the brickwork. What is this followed by?
4 In earth retaining walls where could the attacking sulphate be derived from?
5 Sulphate attack may show differential movement between the faces of brickwork. What is the consequent bowing of the wall a means of relieving?
6 As sulphate attack proceeds what appearance do the mortar joints have?
7 How can you distinguish between moisture expansions of bricks and sulphate attack?
8 What three materials must be present simultaneously for sulphate attack?
9 Which of these two mortars is more liable to sulphate attack?

$$1 : \tfrac{1}{2} : 4\tfrac{1}{2} \quad \text{or} \quad 1 : 1 : 6$$

State one possible reason why.
10 What type of brick should be used for retaining walls?
11 List five design precautions to be taken in detailing a free standing wall.
12 What type of brick and mortar should be used on earth retaining walls?

3 DEWATERING

In order to understand the effect of water on different soils it is necessary to think about the nature of soil. Soils are primarily made up of large numbers of small mineral particles. Between each particle of soil there are small gaps which may be filled with air or water. The size of these gaps will vary from soil to soil, and the ratio of these voids and the size of the particles is called the **void ratio**. It is easy to see why the void ratio of a particular soil will affect its behaviour under load.

Of course, it is quite possible that these voids will be filled by water or that water will pass through them. The ease with which this may happen is called the **degree of permeability**. Those soils with relatively large voids have higher degrees of permeability, and will also therefore be easier to dewater. These soils (like sands and gravels) are called **non-cohesive** soils and are usually free-draining. Soils with smaller voids which tend to hold water are called **cohesive** soils. These are basically clays.

As we have seen water affects the bearing capacity and behaviour of the soil. This has to be taken into account especially when designing and building substructures.

Sub-surface water is the term given to *all* water beneath the earth's surface – most of which comes from rainfall. When the rain seeps through voids it is called **free water** but eventually it will reach a point where all the voids are already filled – where the water is **held**.

Thus, there are two distinct levels in sub-surface water: where there is free water, and where the water is held and becomes **groundwater**. The transition of free water to ground water is called the **groundwater table** (sometimes the water table or phreatic table.)

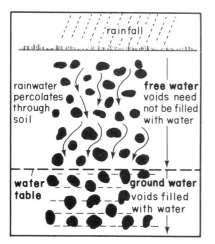

Problems with design and construction occur when the work is below the groundwater table: *all work done below the groundwater table is like working below the water line of a river*. In this course we are concerned only with constructional problems caused by groundwater.

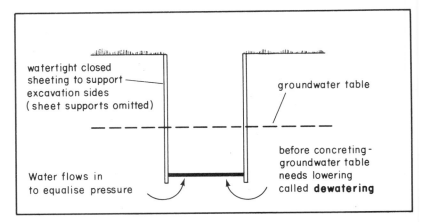

Water flow can be due to pressure, the water flowing from high to low pressure until there is a state of equilibrium. If the water table is lowered by pumping the effect will vary according to type of soil.

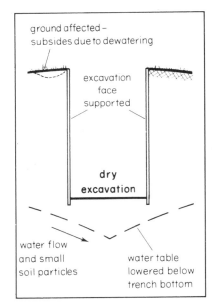

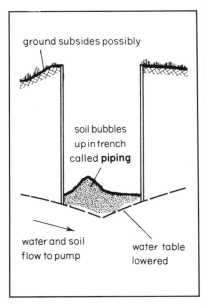

The water carries with it small particles of soil particularly in non-cohesive soils, and this can cause **piping** in the excavation.

Review

1   State the two problems for design and construction of buildings posed by the pressure of groundwater.
2   Define the term 'sub-surface water'.
3   Describe the difficulties of excavation below groundwater table.
4   Define the term 'de-watering'.

WHAT YOU NEED TO KNOW

What information do you require before you can start de-watering? When routine soil investigation reveals the possible need for a de-watering system, additional field work is required to determine the most suitable system of de-watering and to establish design criteria for the system. The most important information which should be obtained is:

1 Level of groundwater and possible fluctuations.
2 Proximity of sources of free water (e.g. pond).
3 Detailed soil profile, including classification and grain size distribution of soils found below the groundwater table.
4 In place permeability of soils below water table.
5 Effect of de-watering on proposed structure and if applicable, adjoining structures.
6 Anticipated water flow in litres/second.

It should always be remembered that the groundwater table is subject to seasonal and other fluctuations. If the site is near a free body of water, variations in the groundwater level are likely to be small. The proximity of a free body of water will normally have considerable influence on the capacity of pumping plant required.

It is particularly important to know if certain strata of soils are continuous over the full area of the site. The presence of a continuous clay or silt seam a few centimetres thick just above the proposed level of pumping in sandy soil has a very great influence on the efficiency of the de-watering system.

Field investigations do not provide a clear cut answer as to the most suitable method of de-watering. The laboratory test results vary too much from the *in situ* values; consequently the permeability of the soil should be estimated in the field by means of a pumping or seepage test.

Review

1 At what stage should groundwater problems be investigated?
2 List six pieces of information required before starting de-watering.
3 State one effect which will alter the groundwater level.
4 What will have a considerable affect on the choice of pumping?
5 State the suggested method for obtaining accurately the permeability of the soil.

DE-WATERING METHODS

The method selected for de-watering will basically be determined by

1   water flow capacities
2   permeability of soil
3   method of construction to be adopted
4   pumping head.

Types of pump available are:

Hand lift diaphragm – suitable for intermittent pumping in small quantities.

Motor-driven diaphragm – can deal with limited quantities of sand/gravels. Its capacity varies but the average capacity with 75 mm suction pipe approaches 1000 litres/hour. A larger suction pipe diameter increases the capacity.

Pneumatic sump pump – can deal with sand, silt in limited quantities. Its capacity at 3 metre head approaches 55 000 litres/hour.

Self priming pump – steady pumping for fairly clean water. Sand and silt in large quantities may cause inefficiency.

Rotary displacement – can deal with considerable quantities of sand and silt. With a 75 mm suction pipe, the capacity approaches 34 000 litres/hour.

Submersible pump – electrically operated and very reliable, carries good pumping head and capacity. Pumping head approaching 40 metres with a capacity 12 000 litres/hour.

Left self-priming pump, *right* motor driven diaphragm, *below* two types of submersible pumps.

**METHOD OF
SUMP PUMPING**

1   Form a sump below the general level of excavation.
2   Select the pump required to cope with water flow, pumping head and soil type.
3   Pump until groundwater table lowered. Pumping may be intermittent or continuous.

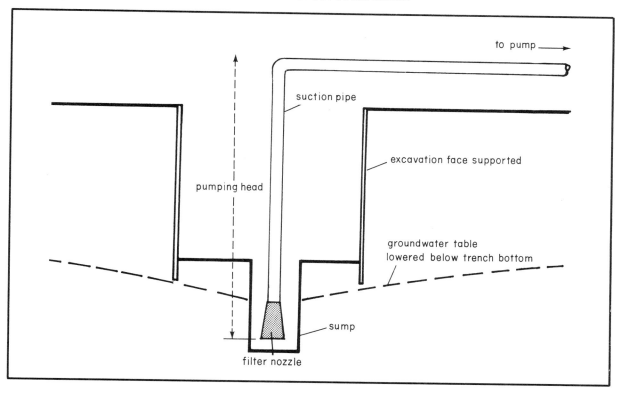

The diagram shows the use of a suction pump such as a self-priming pump. The alternative is to use a submersible pump for better reliability. One disadvantage of sump pumping is that the water flow is toward the excavation, so care is needed to avoid excavation collapse.

Review

1   What determines the method of de-watering?
2   List the types of pumps available for use in sump pumping.
3   Select the type of pump(s) suitable for the following conditions,
    a) limited quantities of sand, 11 000 litres/hour flow anticipated
    b) continuous pumping with clean water, 12 000 litres/hour flow anticipated.
4   Finally try your hand at sketching and describing the sump pumping method to be adopted for a house foundation in sandy clay. The excavation depth is 1·5 metres.

Activity

You will need the following information for this work sheet.

1 Site layout 75/5/1
2 Bore hole logs 1 and 2
3 Construction programme
4 Trade literature on pump

Using the information available it is necessary to lower the water table below excavation depth (this is 4.5 metres). Select the type of pump for sump pumping with the following anticipated water flow 10 000 litres/hour. Indicate on the site layout the positions of sumps and method of de-watering to coincide with the construction programme.

(See answer on page 209)

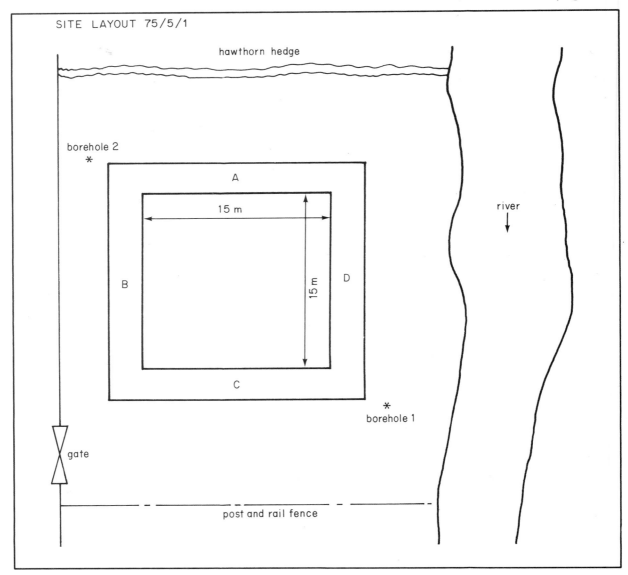

SITE LAYOUT 75/5/1

hawthorn hedge

borehole 2

A

15 m

15 m

river

B

D

C

borehole 1

gate

post and rail fence

BORE HOLE LOG

bore hole   1
type of boring   shell and auger
date   23 Sept

project   offices
bore hole dia.   150 mm
ground level   100.00
water table level   96.50

field classification	sample	depth	BH	depth	OD	remarks
soft brown silt and clay	1 d	1.5			100.00	
soft brown clay	2 d	3.5	−x−x −x−x	3.00 4.00	97.00 96.00	water table 96.50
brown clay with pebbles	3 u 4 d	5.00 6.00		7.00	93.00	
stiff to hard clay	5 u 6 d	7.50 8.00	−o−x −o−x −o−x −o−x	10.00	90.00	

key
d disturbed sample
u undisturbed sample

end of bore hole

BORE HOLE LOG

bore hole   2
type of boring   shell and auger
date   23 Sept

project   offices
bore hole dia.   150 mm
ground level   100.00
water table level   96.50

field classification	sample	depth	BH.	depth	OD.	remarks
soft brown silt and clay	1 o	1.5			100.00	
				3.00	97	water table 96.50
soft brown clay	2 d	3.5	−x−x −x−x −x−x −x−x	5.00	95	
brown clay with pebbles	3 u 4 d	6.00 6.50		7.00	93	
stiff to hard clay	5 u 6 d	7.50 8.00	−o−x −o−x −o−x −o−x	10	90.00	

key
d disturbed sample
u undisturbed sample

end of bore hole

CONSTRUCTION PROGRAMME

operation	contract weeks										remarks
	1	2	3	4	5	6	7	8	9	10	
Excavate to reduced level	▨	▨									
Excavate trenches A			▨								
B				▨							
C					▨						
D						▨					
Concrete trenches A				▨							
B					▨						
C						▨					
D							▨				

4 SHORING

THE PROBLEM A shop keeper approaches you as a builder, wanting the shop window enlarged. The proposed alteration and enlargement is shown in elevation.

enlarge to accommodate shop window

He leaves the elevation drawing with you and arrangements are made for a site visit.

1   *What points or observations will you make during this site visit? List four observations you will make.*
2   *What would happen if you cut out the brickwork as shown on this diagram?*

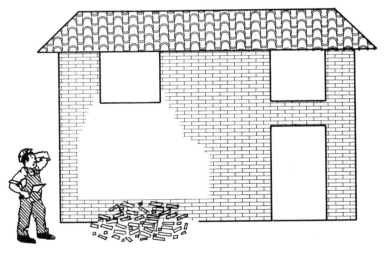

Brickwork will be self-supporting if the wall is raked back at 60°. The area of brickwork supported by a lintel is a triangular shape with 60° angles.

By now you should have learnt the error of your ways. You do not just cut out a hole or form an opening in a wall without first supporting the wall above. This is called **shoring**.

Observe the diagram. We know the effect of just cutting a hole or opening in a wall – the brickwork will collapse around us.

What effect will this have on the first floor windows? Both windows will distort. Now to add to your problems you have a heap of bricks and two useless window frames.

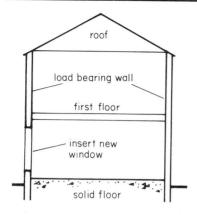

We have only considered the wall, but what if the wall is a **loadbearing wall**?

In addition to the brickwork triangle to be supported we have also to support the floor and roof loads. From the site investigation we know if the wall is loadbearing, that is, it carries roof and floor loads. Over the proposed opening we must temporarily support the floor and roof.

WHAT TO DO

How might you support the roof and floor loads over the opening?

To take the roof and floor load we must insert a temporary shore constructed of wood or tubular steel, sufficiently strong and stable to carry the floor and roof loadings. This temporary support transmits the dead and live loads on the roof and first floor to the ground. Ensure that the loads transmitted do not exceed the safe bearing capacity of the ground.

This temporary supporting system is called the **floor support**. The following components will be involved.

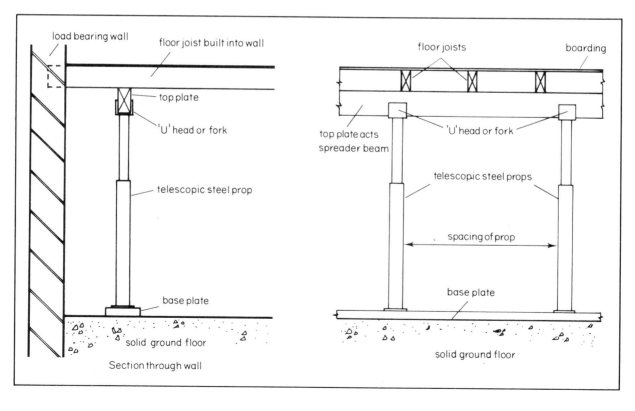

Review

Answer the following questions.

1 What is the function of the top plate?
2 What two factors determine the spacing of the props?
3 What is the function of the base plate?

Now, the floor support system is in position, but is it stable? For vertical loads yes, but not for lateral thrust. The floor support system needs **bracing** to make it rigid and resist horizontal thrusts.

Can you label the components a, b and c on the next drawing?

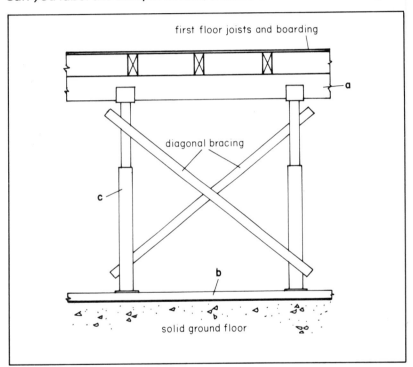

first floor joists and boarding

a

diagonal bracing

c

b

solid ground floor

The roof and floor loads are supported now by the floor support. The next step is to support the brickwork and windows in the wall. This is called **dead shoring**.

WHAT IS THE SEQUENCE OF OPERATION?

From the site investigation information was obtained about heavy traffic and pedestrians. The dead shoring will go on to the path or possibly the road. This will mean getting police permission and signs being erected. Hoardings will also be considered with a temporary walkway for pedestrians, and this will involve local authority permission and licences.

The dead shore consists of three basic components

1 needle
2 dead shore
3 base plate.

The needle carries the wall brickwork load and is positioned through the wall. This load is transmitted to the dead shore. The dead shore acts as a column supporting the needle, transmitting the dead load to the base plate. The base plate transmits and spreads the load over the ground, to avoid settlement.

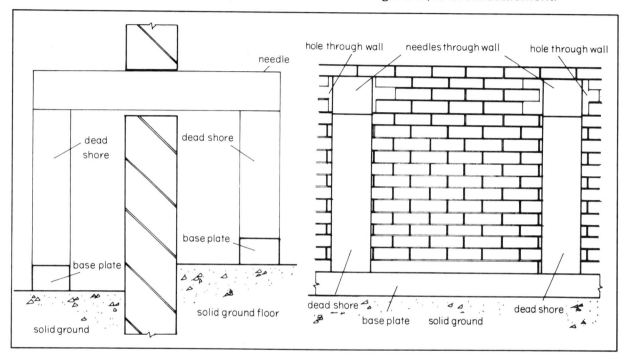

Consider carefully the elevation drawing above. Is the dead shore system stable? Like the floor support system it is stable for vertical loads but not for horizontal loads or thrusts. Bracing is again used to ensure that it is completely stable. If timber is used in the dead shore system then the bracing may well be timber boards. Alternatively if adjustable steel props, with a universal steel beam as the needle, are used then the bracing will be tubular steel scaffolding.

The whole assembly still needs to be made solid. Usually folding wedges are used to ensure a tight fit between the wall and the needle. If the material used is timber, the needle and the dead shore are nailed together with wrought-iron dogs. The dead shore is nailed to the base plate with iron dogs too.

We are now in a position to undertake the work of enlarging the shop front for the shopkeeper. And the best of luck.

Review

Can you label all the components *a* to *f* shown in the drawing?

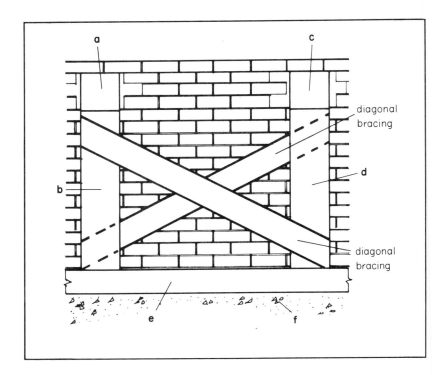

5 RAKING SHORES

THE NEED FOR RAKING SHORING

The strength of a wall is affected by many things – the strength of the bricks, of the mortar, and of the bond between the bricks and mortar. But the most dangerous design fault is what is called an inadequate **slenderness ratio**.

The slenderness ratio is determined by the formula

$$\text{slenderness ratio} = \frac{\text{height of wall}}{\text{wall thickness}}$$

If this ratio is too large the wall may be unable to sustain the weight of the floors. This overloading causes bulging thereby making the building dangerous to the occupants and the public.

Raking shores will give support to a bulging wall. They may be required to provide temporary support to a wall which has become defective and in danger of collapsing, or as a precautionary measure to ensure adequate support to the existing structure during alteration work.

If the following diagram were overloaded at the point indicated, this is what would happen to the wall.

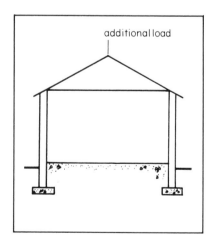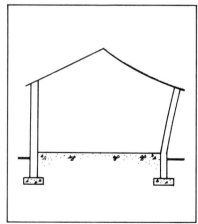

How would you temporarily support this leaning wall? You could prop it up using a raking shore. If the raking shore is to support the **thrust** adequately we must consider

1 force of thrust
2 type of material to be used for raking shore system
3 requirement of a solid base at ground.

The centre line of the raking shore must correspond with the **centre line of the thrust**.

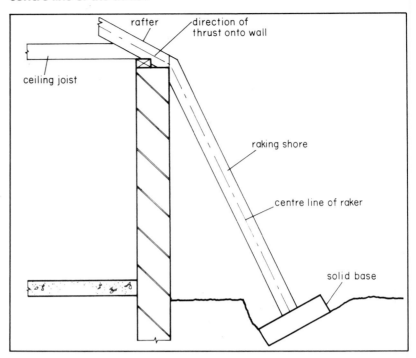

WHERE TO PLACE THEM

Now assume that a three storey building had developed a bulge as shown on the diagram below.

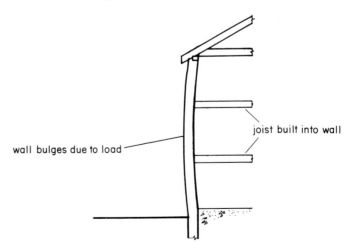

What are your proposals for temporarily supporting the wall? Where would you place the raking shores? How can we ensure a satisfactory connection of the raking shore to the wall, to ensure adequate support?

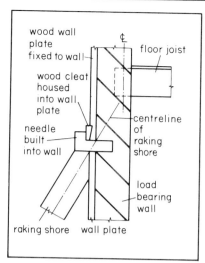

wood wall plate fixed to wall

floor joist

wood cleat housed into wall plate

needle built into wall

centreline of raking shore

load bearing wall

raking shore   wall plate

A wall plate is secured to the wall with wrought iron wall hooks at approximately 2·5 m between the centres. To connect the raking shore and the wall plate a **needle** and **cleat** is used. The needle goes through the wall plate and into the wall at usually half brick depth.

The raking shore is notched to take the needle. *Can you name the unlabelled components a, b, c and d?*

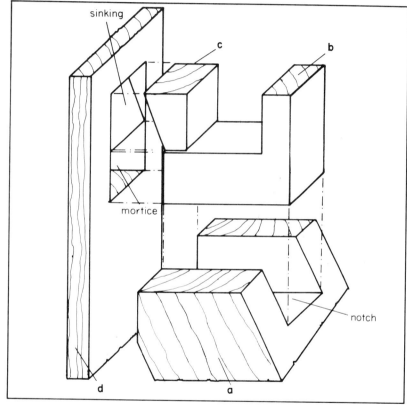

sinking

mortice

notch

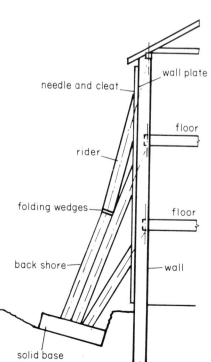

needle and cleat

wall plate

floor

rider

floor

folding wedges

back shore

wall

solid base

The length of the top raking shore may be too long. The top raker is then split into

    a) **back shore**
    b) **rider**.

To ensure lateral stability the raking shores are braced. At the base, hoop iron binding is an alternative to timber boarding as a bracing. This system is called a compound raking shore system, (as opposed to the single raking shore system).

Because the raking shore system has to resist the wall thrust, care must be taken to ensure that the shore spreads the load over the ground to avoid settlement or sinking. The rake or angle of the raking shore should be 60° to 75° to the horizontal.

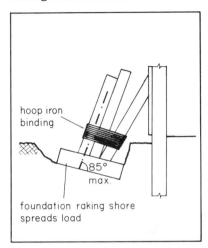
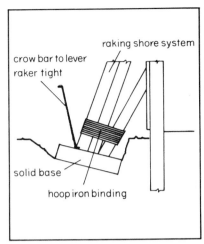

When the system is assembled, it is essential that a tight fit to the wall is achieved, or further bulging will occur. The tightening of the system is done by wedging at the base and finally nailing the members with dog irons.

Review

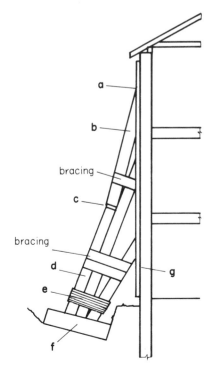

1 State two reasons for raking shoring.
2 Sketch a typical detail showing how the raker is connected to the wall. Label all the components.
3 At what angles to the horizontal should the rakers be positioned?
4 Sketch the position of the 'rider' and state when it is used.
5 What angle should the base plate be to the top raker?
6 Write down the labels a to f for the drawing shown.

6 BASEMENTS

Read through the Building Regulations, Part A4(2).

A4(2) In these regulations–

(a) BASEMENT STOREY (except in Part E) means a storey which is below the ground storey; or, if there is no ground storey, means a storey the floor of which is situated at such a level or levels that some point on its perimeter is below the level of the finished surface of the ground adjoining the building in the vicinity of that point;

GROUND STOREY (except in Part E) means a storey the floor of which is situated at such a level or levels that any given point on its perimeter is at or about but not below the level of the finished surface of the ground adjoining the building in the vicinity of that point; or, if there are two or more such storeys, means the higher or highest of these;

SINGLE STOREY BUILDING means a building consisting of a ground storey only; and

UPPER STOREY means any storey other than a basement storey or ground storey; and

(b) unless the context otherwise requires, wherever these regulations describe a building or part by reference to a number of storeys, that number does not include basement storeys.

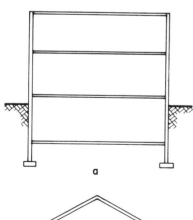

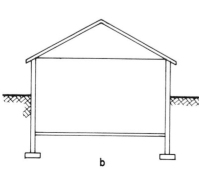

Now answer the following questions.

1 Define a basement.
2 On which of these diagrams is there a basement storey?

a

b

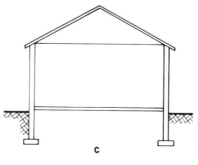

c

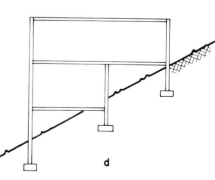

d

OVER EXCAVATION

If we excavate a basement as shown, what will happen to the face of the excavation?

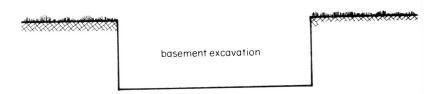

basement excavation

The face of the excavation will collapse if unsupported. The excavation face then will need supporting, unless we **over excavate**. *What is the angle called where the soil is self supporting?*

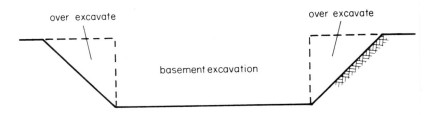

over excavate   over excavate

basement excavation

This method of excavating a basement, where the soil needs no support, is called **over excavation**, or the 'battered excavation method'.

Sometimes it is not possible to use the over excavation method of excavating a basement. *Can you think of such an occasion?*

Consider a traditional house foundation trench. *How are the excavation faces supported temporarily? What would be a satisfactory temporary support for this excavation?*

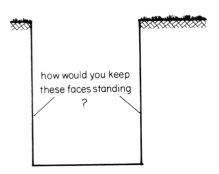

how would you keep these faces standing ?

THE DUMPLING METHOD

It might be possible to use the following method of excavating a basement. Excavate the perimeter wall like a trench, build the basement wall. When the basement walls are capable of acting as a retaining wall then excavate the middle of the basement.

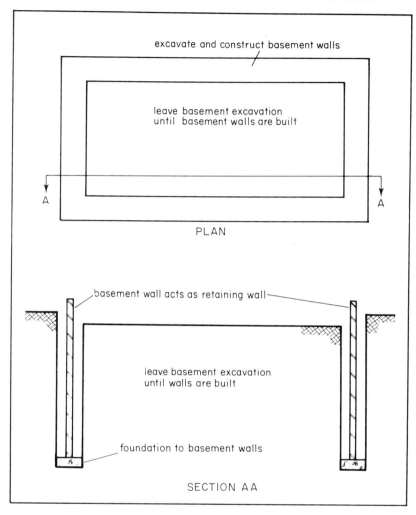

excavate and construct basement walls

leave basement excavation until basement walls are built

A   A

PLAN

basement wall acts as retaining wall

leave basement excavation until walls are built

foundation to basement walls

SECTION A A

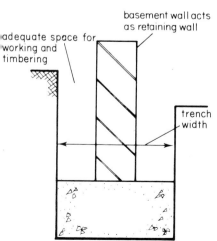

adequate space for working and timbering

basement wall acts as retaining wall

trench width

The width of the trench must be wide enough to accommodate planking and strutting. It must have sufficient room for workmen and the basement wall. The room required for the workmen is called working space. The deeper the trench the more width is required to accommodate working space. This method of excavating a basement is called the **dumpling method**, or 'trench and dumpling method'.

Leaving centre excavation until after the basement wall has been constructed restricts the choice of excavation machine to some extent.

THE DIAPHRAGM METHOD

Another method similar to the dumpling method of excavation is called the **diaphragm wall** technique. With this method the basement wall is usually made out of reinforced concrete. A section of the wall is excavated, filled with a thixotropic suspension fluid such as the clay called bentonite. When the bentonite fluid is in the excavation there is no need to support the excavation faces. Mild steel reinforcing cages are introduced into the bentonite. A good quality concrete is then poured into the section of excavation containing the bentonite. The bentonite suspension fluid is displaced by the concrete and collected for further use.

So far we have discussed three techniques for excavating basements, namely

1 over excavation
2 dumpling (trench and dumpling)
3 diaphragm walling

DEEP EXCAVATION

There is another method of excavating the basement which adopts techniques used in raking shoring.

The basement is excavated to the minimum size. Planking and strutting is introduced to support the excavation face. Instead of using struts in the dumpling technique the planking and strutting is propped from the bottom of the basement with shores.

This method is called deep excavation.

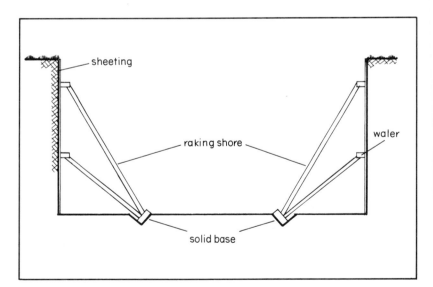

Review

List the four methods of excavating a basement, and explain how they differ.

EXCAVATING PLANT

PHEW! CAN'T WE USE A MACHINE TO EXCAVATE THIS BASEMENT?

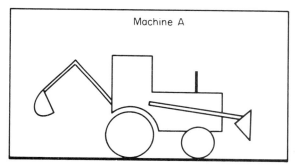

We will consider plant needed for two methods, deep excavation and dumpling, in detail.

There are several factors which influence the choice of excavating equipment, namely

> ground condition, size of basement, depth of basement, method of excavating, construction sequence.

Often the building site will be like a quagmire at this substructure stage.

During the site layout, pre-planning will save time and money if correctly laid traffic routes are adhered to. Temporary roads are well kept. Sometimes the ground condition will dictate the type of excavating machine to be used. *On poor ground conditions which machine would be more suitable?*

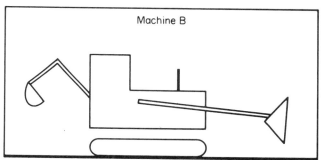

Machine A has the load transmitted through the tyre giving a heavy point load, whereas machine B spreads the load across the length of the machine through tracks. The first classification of excavating machines is into **tyred** (or wheeled) and **tracked**. These machines are called excavator/loaders, and are common plant items on a building site.

The machine here is hydraulically operated. This gives the machine operator greater control of the excavator arm and more power.

The other type of excavator is the wire or rope controlled excavator. These have larger **reach** or depth of digging than the hydraulically operated machine.

From the table opposite you can note the differences in depth of dig and reach of the various machines. This limitation of the excavator, its dig depth and reach, is important in considering the factors which affect the choice of equipment.

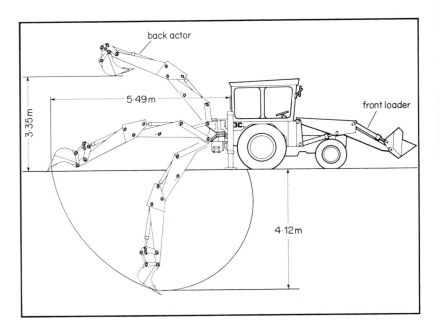

Table of machines and their reach/dig capacity

Type of excavator	Manufacturer	Backhoe/excavator		
		Digging depth	Reach at ground level	Max. digging force
		m	*m*	*kg*
OFFSET EXCAVATOR	J. C. Bamford 3D	4·68	6·17	4699
	Ford 4550	4·11	5·35	4264
	Hymac 570	4·80	6·60	9800
	Massey Ferguson MF50B	4·70	5·16	4050
WHEELED HYDRAULIC EXCAVATORS	Hymac 596	5·64	8·92	6395
	Poclain GY120	6·75	10·50	*N/A*
TRACKED HYDRAULIC EXCAVATOR	JCB 808	7·01	10·21	13119
	Caterpillar 235	8·1	11·9	23100
	Hymac 1290	8·64	12·65	10206
	(Dragline attached)			
TRACKED ROPED OPERATED EXCAVATORS	Babcock & Wilcox 101MB	9·2	16·8	17526
	Bucyrus Erie 88B	10·67	25·4	23920
	Priestman Lion 350D	6·4	17·37	9444

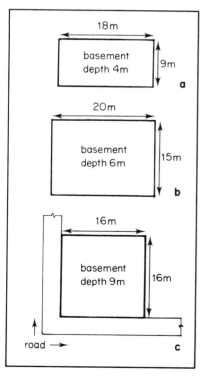

We can now consider the relation of size and depth of basement to choice of machine.

Exercise

Try and work it out yourself. *Using the table of machines and their reach/dig capacity, select suitable excavators nearest to the basement sizes for* **a** *and* **b**.

For those two examples we have been allowed to move the excavators all around the basement perimeters. This is often not possible in practice.

Select a suitable excavator again. *How would the machine excavate the basement in* **c**, *i.e. where will the plant be positioned?*

(See answer on page 209)

From the exercise it is apparent *a*) that choice of machine can affect the construction sequence and the method of excavating the basement; *b*) that rope excavators are suitable for larger deeper basements than their counterpart hydraulically operated excavators.

Review

To summarise this section complete these questions.

1 List the five factors affecting the choice of excavating equipment.
2 Excavators are classified according to method of mobility and operation. State the two methods of mobility. State the two methods of operation.

Exercise

We are now in a position to consider this project. A commercial development comprising of ground floor shops with some basement area is planned. The rest of the four storey structure is devoted to offices.

Draw up a proposed site layout.

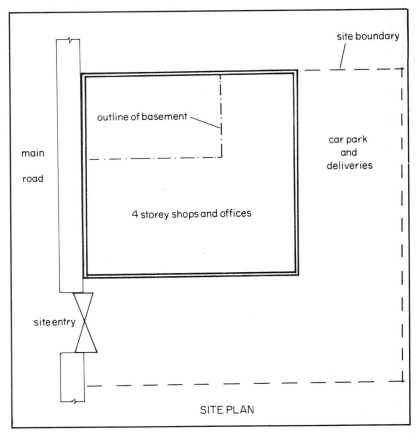

SITE PLAN

We have obtained the following soil information. The soil is firm to stiff clay and the water table is 6·00 m below ground level approximately. *Consider the choice of excavation method.* There are four methods, namely over excavate, dumpling, deep excavating, diaphragm walling. Which of those four methods is automatically omitted?

Describe the method of excavating a basement a) using the dumpling method and b) using the deep excavation method.

We do not have to consider temporarily lowering the groundwater table, so de-watering is not a constructional problem.

In the site layout, for the site security and the public's safety, a hoarding has been erected along the main road boundary.

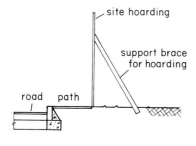

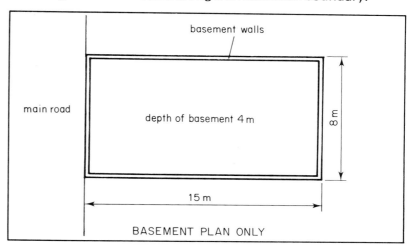

How will we support the hoardings if we use the deep excavation method?

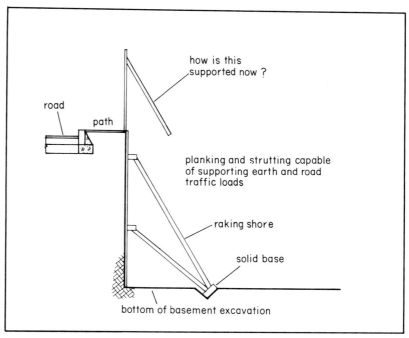

It is not an impossible task but has to be considered. For this reason the site manager has decided to excavate the basement using the dumpling technique.

This is the site manager's proposed sequence of excavation.

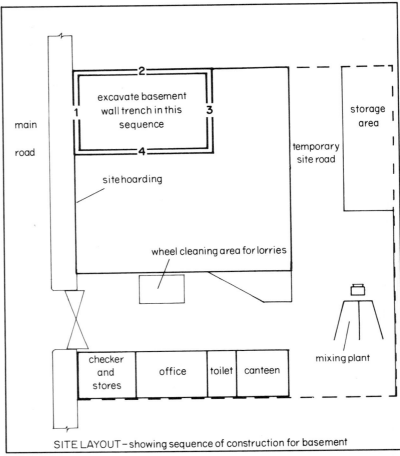

The excavation sequence is indicated above; notice the wheel cleaning area for lorries carting away basement excavation off site. *What machine will be used for excavating?* The machines are classified as tracked (crawler) or tyre (wheel).

The ground excavation is timed for the summer months so the site manager decides to use a tyred excavator, giving him greater mobility. The next decision is whether to use a wire or hydraulic excavator. This will depend on reach/digging depth and the excavating force required. The soil is stiff to firm clay so the site manager has decided to use the more powerful excavating machine, namely hydraulically operated rather than the wire operated excavator. *Use the table on page 49 and select a suitable hydraulically operated excavator.*

The sequence of excavation, the type of excavator, has been decided. Now we must consider the support to the trench face. Some soils are self supporting in certain conditions, others need supporting immediately after excavation. Clay soil in this excavation will require only open timbering unless there are other factors to be considered. These factors are

1 poor ground conditions
2 excavation next to highway
3 excavation face exposed for long periods.

The temporary supporting of the trench excavation is covered by the Construction (General Provisions) Regulations. Basically the Regulations state that an adequate supply of timbering is to be provided to prevent danger of collapse, which may endanger the lives of those working in the excavation. The excavation face and, if necessary, the planking and strutting must be inspected once every day, or after adverse weather. A record of the examination of the excavation face and timbering must be recorded in the Register for Inspection and Examinations of Excavations weekly. If the trench is more than two metres deep a barrier must be erected to prevent people and machines falling into the excavation. Finally there must be a means of safe access and egress from the excavation.

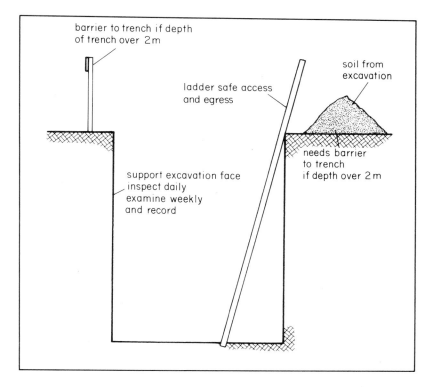

Planking and strutting materials are usually timber or steel. Two factors will determine the material: strength required, and the cost.

Where would open timbering and where would closed timbering be required for the basement excavation on p 50?

Review

1 What factors are also considered when supporting trench or excavation faces, where only open planking and strutting is needed?
2 What is the inspection period for planking and strutting according to the Construction Regulations?
3 If the excavation is over two metres deep, what must be provided, apart from safe access and egress?
4 Now with your expert knowledge you can help the builder decide on the method of excavation, safety precautions and selection of the excavating plant to suit the builder's programme. Study carefully the contract drawings, grid plan and section through basement in conjunction with the works programme.

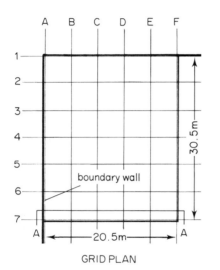

GRID PLAN

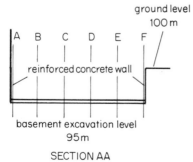

SECTION AA

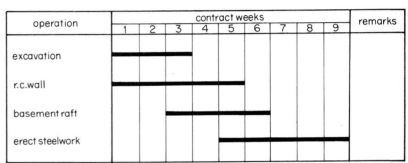

First, decide on the number of machines required to keep to programme; next, choose a method of excavating and note your reasons for your choice; finally, select a suitable excavating machine from your notes.

The excavator's output is 14 m³ per hour.

Summarise your work into a report format under these headings.

a) Method of working and reasons.
b) The plant requirements.
c) The safety precautions to be adopted and the requirements of the Construction (General Provisions) Regulations.

(See answer on page 210)

7 STRUCTURAL STEEL

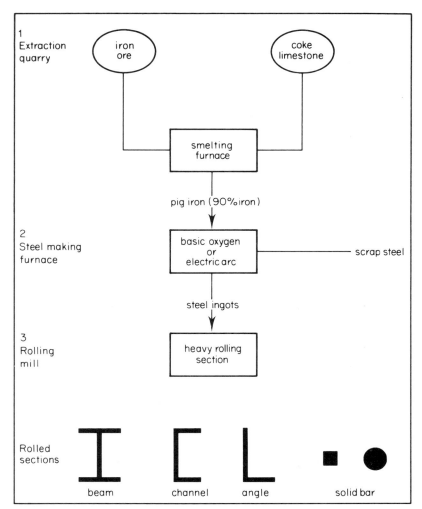

1
Extraction
quarry

iron
ore

coke
limestone

smelting
furnace

pig iron (90% iron)

2
Steel making
furnace

basic oxygen
or
electric arc

scrap steel

steel ingots

3
Rolling
mill

heavy rolling
section

Rolled
sections

beam   channel   angle   solid bar

Steel is refined iron with additions of chemicals to improve various inherent properties. Iron ore from the quarries is smelted with coke and limestone. This smelting process produces iron called pig iron. It is about 90 per cent pure and contains various other elements such as carbon, manganese, silicon, phosphorus.

This pig iron is then further refined in a furnace. There are three major smelting processes – open hearth, electric arc, or basic oxygen. During the refining process the pig iron is heated with scrap steel.

Impurities are eliminated as slag and the refined steel is cast into ingots for rolling into various shapes. The chemical composition is strictly controlled in accordance with British Standard 4360. The steel is then classified into grades. For structural steel there are three grades of steel, Grade 43, Grade 50, and Grade 55. According to BS 4360 certain maximum percentages of certain elements are defined, for example the content of carbon, silicon manganese, niobium sulphur and phosphorus.

The ingots are then reheated in soaking pits, rolled down and the rough ends cropped off. The steel is now in the form of 'blooms'. The bloom is cut to length and allowed to cool and after the product examination according to BS 4360, it enters the rolling mill. (For further reading see *'Architect's Journal' Handbook of Building Structure* ed. Alan Hodgkinson, Architect Press.)

Rolled sections are available according to BS 4 *Structural Steel Sections*, Part 1 *Hot Rolled Sections*. This British Standard covers universal beams, universal columns, channels, and angles. It specifies the size of the rolled section and permitted deviations or 'tolerance' from the specified shape. It also lays down minimum thickness for the web and flange. It is usual not to specify the thickness of these two components but to specify the weight of steel per section. A **universal beam** is thus labelled like this: $914 \times 305 \times 253$ kg/m.

Universal columns are similarly designated. BS 4 states the overall size of this square rolled section, the thickness of the web, flanges, and also the weight of steel section per metre length.

This universal column is a $254 \times 254 \times 73$ kg/m UC

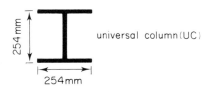

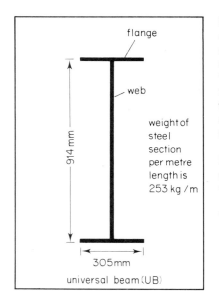

What size of beam or column is required? The size of structural members is the domain of the structural engineer. The consultant structural engineer will calculate the maximum bending moment, shear force, and actual deflection of the beam. His selection of the size of beam capable of safely carrying imposed and dead loads depends on the moment of inertia of the steel section. For the rolled sizes this information regarding moment of inertia and other structural properties is published in the *Handbook of Structural Steelwork*. This handbook is published jointly by the British Constructional Steelwork Association and the Constructional Steel Research and Development Organisation. For mere mortals this handbook also includes load/span tables, like the floor joist tables published in the Building Regulations. What has to be calculated is the load; that is, the safe distributed load in kilonewtons (kN). (For further reading see *Structural Steelwork for Students* by L. V. Leech published by Newnes-Butterworth, 1972.)

CONSTRUCTING THE STEEL FRAME

How is the structural steel frame connected together?

It is common practice to bolt the beams to the column. This means the column needs to have a cleat welded to it marking the position of the beam connection. The cleat serves as a fixing cleat and location cleat during erection. The site connection will be bolted.

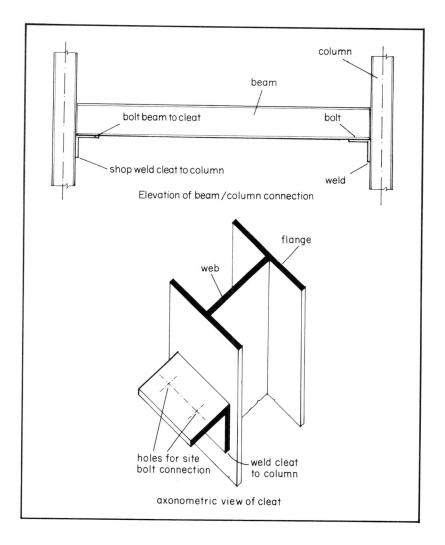

column

beam

bolt beam to cleat

bolt

shop weld cleat to column

weld

Elevation of beam/column connection

flange

web

holes for site bolt connection

weld cleat to column

axonometric view of cleat

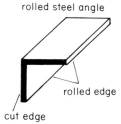

rolled steel angle

rolled edge

cut edge

BS 449 specifies the minimum cut and rolled edge clearances depending on the diameter of bolt used in the site connection.

Bolt and rivet spacing (extract from BS 449)

Diameter of hole mm	30	24	20	18	16
Cut edge clearance mm	50	38	30	28	26
Rolled edge clearance mm	44	32	28	26	24

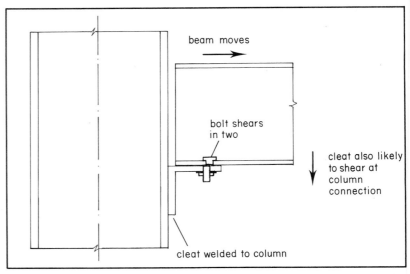

beam moves

bolt shears
in two

cleat also likely
to shear at
column
connection

cleat welded to column

BS 449 also gives suitable cleat sizes. The cleat, and especially the bolt, will have to resist the shearing force created by the loaded beam. The cleat is also likely to shear at the column connection.

BS 449 can be used to size the cleat according to the type of bolt to be used. There are several types of bolts used; we will be concerned with the use of **black bolts**. Black bolts require a clearance around the hole because they are not machined accurately. Usually the hole is drilled 2 mm wider than the diameter of the black bolt. This is called a clearance hole.

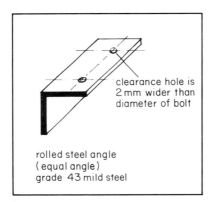

clearance hole is
2 mm wider than
diameter of bolt

rolled steel angle
(equal angle)
grade 43 mild steel

The black bolt has sufficient strength to resist the shear imposed. Using black bolts we have to have a clearance hole. For a 16 mm diameter bolt, the clearance hole is 16+2=18 mm diameter. There is also a tolerance, usually of 6 mm, on the length of the beam. We have to take this into consideration in choosing the cleat size. Notice the beam end is a cut edge and not a rolled edge.

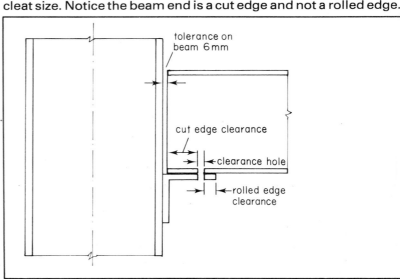

tolerance on
beam 6 mm

cut edge clearance

clearance hole

rolled edge
clearance

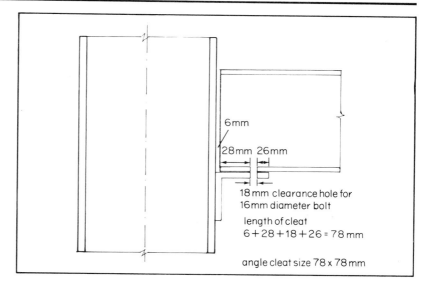

6mm

28mm 26mm

18mm clearance hole for
16mm diameter bolt

length of cleat
6 + 28 + 18 + 26 = 78 mm

angle cleat size 78 x 78 mm

The size of the cleat above must be at least 78 mm. The sizes of angle sections are published in the British Standard and the BCSA *Handbook of Structural Steelwork*. A suitable cleat size would be 90 mm × 90 mm × 16 kg/m. Of course the structural engineer would calculate the required cleat size and then check to see if it conforms with the edge clearances as outlined above.

BS 449 also specifies the minimum pitch between bolt holes as $2\frac{1}{2}$ times the diameter of the bolt.

Let us build up a cleat connection using a 16 diameter black bolt. The UC size is 152×152×23. It will safely carry the load imposed, but this is too small for erection purposes.

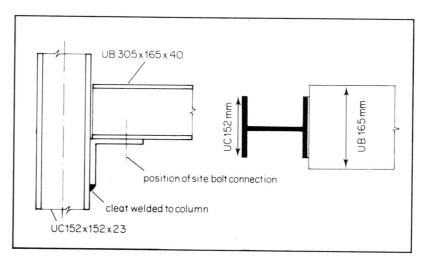

UB 305 x 165 x 40

UC 152 mm

UB 165 mm

position of site bolt connection

cleat welded to column

UC 152 x 152 x 23

It is best to have the beam width equal to or less than the column size. In the case above we would have to select a larger universal column.

Review

Copy and complete the following detail of a beam to column connection. Label it fully and indicate the type of site connection.

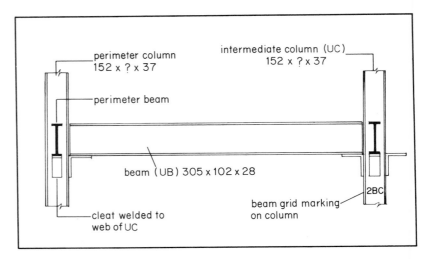

The structural steel frame is connected by angle cleats welded on to the universal column. After the columns are erected on to the pad bases, the steel universal beams can be lifted and placed into the final position. The erection work is made easier by the beam/column referencing to the structural grid. Once the beam is in position the beam/column connection can be completed by site bolting.

CONSTRUCTING THE BASE

Sketch a typical plain concrete pad base suitable for framed structure.

The building load is concentrated on the universal columns section. This is usually too great a concentration of load for the concrete pad to carry.

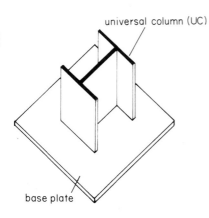

Another component is needed to spread the load concentrated in the universal column. This component is called a **base plate**. The function of the base plate is to spread the building load over a sufficient area of concrete base to avoid overloading the concrete foundation.

The base plate is welded to the universal column. It is important that the universal column is machined dead square to allow for accurate plumbing of the column on site. We need some way of introducing adjustment of the column for plumb and alignment. British Standard PD 6440 *Accuracy in Building* sets down permissible tolerances for various building operations. Steel frames are to be erected to ±15 mm per 30 metres high.

The concrete base will not be finished accurately enough for a deviation of ± 15 mm in 30 metres. The base detail for steel frames uses locating bolts with a tolerance for alignment to the structural grid and packing for vertical alignment.

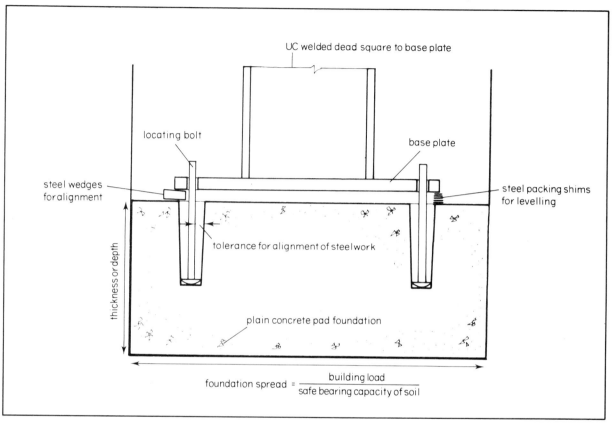

UC welded dead square to base plate

locating bolt

base plate

steel wedges for alignment

steel packing shims for levelling

thickness or depth

tolerance for alignment of steelwork

plain concrete pad foundation

$$\text{foundation spread} = \frac{\text{building load}}{\text{safe bearing capacity of soil}}$$

The setting out operation combines three operations, namely alignment, plumbing, and levelling of column. The plumbing and levelling is done using packers and wedges.

Once the column has been correctly positioned, aligned, plumbed and levelled, the base can be grouted up using neat cement grout.

Review

1   Identify the following sections.

a   b   c   d

2   Sketch and label a standard beam to column connection.
3   Sketch and label a column base detail to a plain concrete pad foundation.

8 REINFORCED CONCRETE

We have dealt with framed buildings using structural steel. Now we will consider framed buildings using reinforced concrete instead of structural steel.

THE CODE OF PRACTICE

The concrete design is governed by British Standard Code of Practice 110, (abbreviated to CP 110).

Plain concrete can offer resistance to compressive loads but needs additional material to withstand the tensile loads. Plain concrete and reinforced concrete are therefore used for different purposes. *Where would reinforcement be needed to resist the tensile load on this beam?*

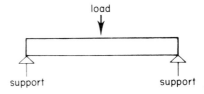

STEEL REINFORCEMENTS

To help the concrete beam mild steel bar is traditionally used as reinforcement. The amount of reinforcement depends on the beam load and span. CP 110 acts as a guide in giving the minimum steel reinforcement requirements.

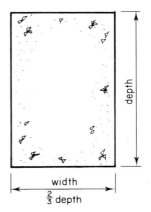

Take a section through a beam. The area of steel must be at least 0.68 per cent of the concrete area. The width of the beam must be $\frac{2}{3}$ of the depth.

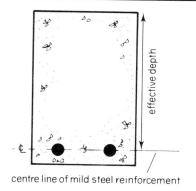

centre line of mild steel reinforcement

Another term introduced by CP 110, is **effective depth**. This is the measurement from the top of the beam to the centre line of the steel bar reinforcement. The effective depth is first obtained using the span/depth formula:

$$\text{Simple span beam} - \text{effective depth} = \frac{\text{Span}}{20}$$

The formula comes from CP 110 and we can use these guides in the Code of Practice to give the beam size and the amount of reinforcement. A structural engineer would also consider the maximum bending moment, the deflection and shear forces.

BEAM DESIGN

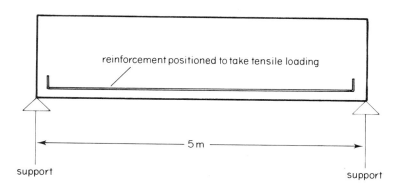

reinforcement positioned to take tensile loading

5 m

support   support

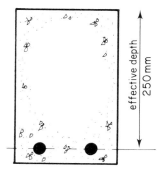

Taking a simply supported beam like the one shown above, the span/depth ratio is

$$\text{Effective depth} = \frac{\text{Span}}{20} = \frac{5000}{20} = 250 \text{ mm}$$

Now consider the number and size of bar reinforcement needed. We have obtained the effective depth. What we need is the concrete beam depth. We must now assume a depth and by trial and error ensure 0·68 per cent of steel to concrete area is maintained. While mild steel plain round bars are usually used, there are also other types of reinforcement bars available. The mild steel plain round bar is manufactured to BS 4449.

Typical sizes of mild steel bar

Diameter size in mm	6	8	10	12	16	20	25
Cross sectional area of bar mm²	28·3	50·3	78·5	113·1	201·1	314·2	490·9

Let us try and put two No. 20 diameter bars into the beam. The area of steel using the table above for two No. 20 diameter bars is

$$2 \times 314 \cdot 2 \;=\; 618 \text{ mm}^2 \text{ steel area.}$$

What will be the actual beam size? The effective depth is 250 mm, half bar is $20 \div 2 = 10$, the cover is say 25 mm, so the beam depth is 285 mm.

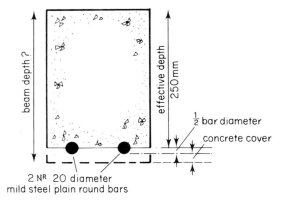

2 NR 20 diameter
mild steel plain round bars

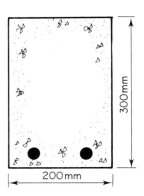

It is preferred that the depth of beam is in increments of 25 mm, therefore, our beam is not 285 mm but 300 mm. The width of the beam is $\frac{2}{3}$ depth, namely 200 mm. Now let us check to see if the steel area is at least 0·68 per cent of concrete area.

Concrete area is $300 \times 200 = 60\,000$ mm²

What is 0·68 per cent of that area?

$$1\% = \frac{60\,000}{100} = 600 \text{ mm}^2$$

$$0·68\% = 0·68 \times 600 = 408 \text{ mm}^2$$

We must have at least 408 mm² of steel bar reinforcement.

The two No. 20 diameter mild steel bars suggested will have a cross sectional area of 618 mm² so this reinforcement will meet the requirements.

That was not so difficult. Why not try this example yourself. Remember the table giving you the size and cross sectional area of bar reinforcement. (See page 63)

Exercise

Calculate size and main bar.

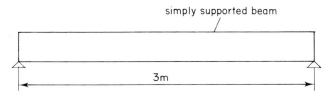

simply supported beam

3m

(See answer on page 211)

The bar reinforcement that you have just calculated is called the **main beam reinforcement**. This will take the tensile load. You will remember that on a loaded beam we have to consider not just bending but shear.

Where would maximum shear occur?

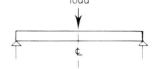

STIRRUPS

Code of Practice 110 gives the following guide to the positioning of steel shear reinforcement. Shear reinforcement must be positioned at the ends since this is where maximum shear occurs.

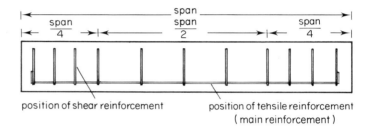

position of shear reinforcement

position of tensile reinforcement (main reinforcement)

The shear reinforcement is spaced at half effective depth for a distance of a quarter span. What does this shear reinforcement look like?

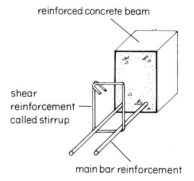

reinforced concrete beam

shear reinforcement called stirrup

main bar reinforcement

So these stirrups are positioned thus:

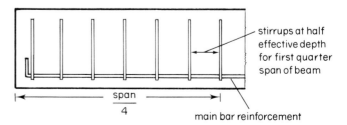

stirrups at half effective depth for first quarter span of beam

main bar reinforcement

$$\frac{span}{4}$$

Part elevation of beam

At the midspan the shear force will be zero for the simply supported beam that we are considering, therefore the stirrups can be spaced further apart at the effective depth spacing. The complete shear reinforcement will be:

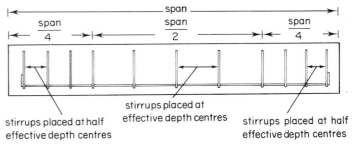

Beam elevation showing shear reinforcement

CONCRETING THE BEAM

What will happen when the beam is concreted?

To avoid the stirrups being displaced during concreting, an additional horizontal bar parallel with the main bar is incorporated to hold the stirrups in position. This is called **top steel**.

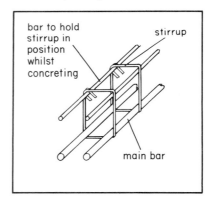

To review, the reinforcement in a typical beam will be:

main bar (to take load);
stirrups (to take shear force);
top steel (to keep stirrup in
position during concreting).

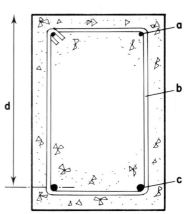

Can you label a, b, c *and* d?

COLUMNS

Let's turn our attention to the column now. Again we can use the guidance of the Code of Practice 110 in determining the size and number of reinforcing bars needed. *What effect will the beam load have on the column?*

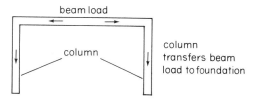

beam load

column

column transfers beam load to foundation

column

column height

width = height / 15

The higher the column the greater the possibility of the column buckling under load. Several factors must then be considered, like the height of column and the cross sectional area of the column. CP 110 states that the height of the column should not be more than 15 times the width. If the column is not square, the minimum side dimensions should be used.

The concrete will need reinforcing in most circumstances. At least 4% of the total concrete area must be made up of steel bar reinforcement.

main column steel bar reinforcement

It is accepted practice to have main column reinforcement in multiples of two, but what should the diameter of the mild steel bar be?

Take for example a column which is three metres high.

$$\text{Using CP 110, minimum width} = \frac{\text{height of column}}{15}$$

$$= \frac{3000}{15} = 200 \text{ mm}$$

The area of steel reinforcement must be 4 per cent of the concrete area.

Concrete area $= 200 \times 200 = 40\,000$ mm²

Therefore 4% $= 1600$ mm².

Bar reinforcement of 1600 mm² cross sectional area is needed.

Remember that main column reinforcement is in multiples of two, so let's try four bars.

$$\text{Cross sectional area per bar} = \frac{\text{Steel area}}{\text{No. of bars}} = \frac{1600}{4} = 400 \text{ mm}$$

Using the table on page 63 the diameter of the bar will be 25 mm.

What will happen to the main column steel when the column is loaded? The tendency is for the main steel to buckle. An additional reinforcing bar is used called a **lateral tie**.

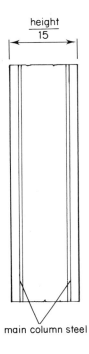

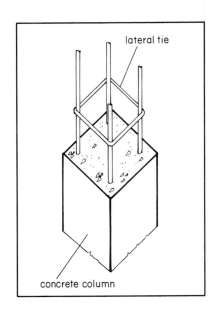

main column steel

The lateral ties are positioned along the main column bars according to the recommendations in CP 110. Lateral ties are positioned at whichever is the greatest of the following:

   1   12 × main bar diameter
   2   Least lateral dimension of column
   3   Every 300 mm

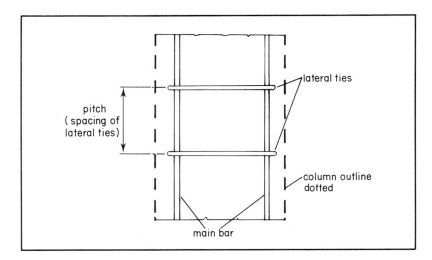

Now we are familiar with the position of reinforcement in beams and columns, we can look at a single storey framed building.

A client wants a framed building in reinforced concrete. The overall building size is 25 metres × 15 metres.

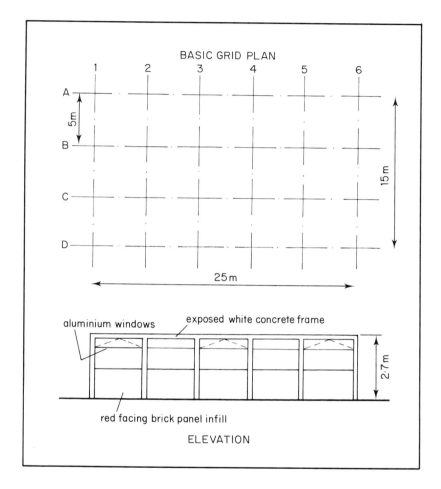

BASIC GRID PLAN

ELEVATION

Prepare the following:

1. The detail of beam B 2/3 in elevation and cross section showing the size and location of the reinforcement.
2. The detail of column B2 in elevation and cross section.

BENDING SCHEDULES

Imagine all the steel reinforcement for that project arriving on an articulated lorry as straight bars. The steel fixer would have to cut and bend the steel to individual shapes to suit the beam or column requirement. To aid the steel fixer there are steel layout drawings and also **bending schedules**. The bending schedules are used by the steel fixer to cut the bar from delivered steel to length, then to bend it to the required shape.

Take for example the beam you have just detailed, beam B2/3. How many different shapes of steel bar are needed? How will you identify these bars?

A different shaped bar will have a different bar marking. This individual bar reference is found on the steel layout drawing as well as on the bending schedule.

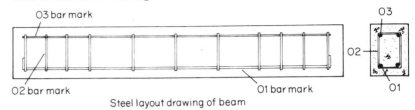

Steel layout drawing of beam

The steel fixer also wants to know

1   what diameter and type of steel to be used
2   what length to cut
3   the shape of the finished reinforcement.

What will happen if the steel fixer cuts the bar to 5·00 m?

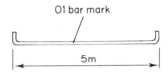

So, the length is the total length before bending.

The shape of finished reinforcement is called the **bending shape**. British Standard gives standard bending shapes as a code number, for example a straight bar is Code 20, whereas a stirrup is Code 60.

What is the diameter of the steel bar mark 01? 16mm diameter mild steel. According to the cross section there are two bar mark 01. The bending schedule for this beam layout drawing will be:

BENDING SCHEDULE					DRAWING No.		
Location	Bar mark	Type & size	No. in member	No of member	Total	Length (metres)	Shape code
Beam B2/3	01	16 ms	2	10	20	5·06	35
	02	12 ms	31	10	310	0.76	60
	03	10 ms	2	10	20	4.90	20

How many similar beams are there? (See the grid plan, page 69.)

Now prepare a bending schedule for the column detail you have prepared, column B2. Do not calculate the length of the lateral ties.

All of that work, thankfully, is the realm of the structural engineer. We must now concentrate on building to his detailed design.

9 CONCRETE MIXING AND PLACING

MIXERS There are three main types of concrete mixers. The **tilting drum** mixer empties concrete from the drum by tilting it. The **reversing drum** mixer has mixing blades which keep the concrete in when mixing, but when reversed force it out. The **pan mixer**, based on the old mortar mill mixing methods, has a horizontal drum with the mixing blade agitating the concrete until it is mixed.

tilting drum mixer

pan mixer

Before we start mixing, ancillary equipment is also needed. We must have storage for cement and aggregate, loading equipment, weighing equipment and a water supply. A typical batching plant on a site could look like this. The mixer is a reversing drum mixer.

Here the cement is stored in a silo. *How else could it be stored? What problems does this alternative method present?*

Mixer size

The batching plant must be able to meet both peak demands and normal week by week requirements. A works programme may for example demand 150m³ in one week for concrete foundations.

Assuming a normal 5 day week a 30m³ capacity would meet peak requirements. The normal working day is 8 hours, however because you have to have time to place and finish the concrete, only 7 to $7\frac{1}{2}$ hours of mixing is available. If there were $7\frac{1}{2}$ hours, a mixer with a capacity of 4m³ per hour would meet the requirements.

Common mixer sizes are 280/200, 350/280, 500/350 and 600/400. Why is each mixer given two capacities? This is because there is a difference between the volume of **dry yield** (made up of the separate constituents) and of the **mixed yield**. The volume of the mixed yield is *smaller*. Thus a mixer with capacity 280/200 means a maximum 280 litres dry yield, and 200 litres mixed yield.

CHOOSING A MIXER

Going back to our example, we need a mixer with maximum capacity of 4m³ per hour, mixed yield. Concrete mixer manufacturers give mixing times for their equipment, but as a guide, a reversible drum mixer takes about $2\frac{1}{2}$ minutes per batch. So we need to calculate the minimum batch size to keep to targets.

$$\text{batch size} = \frac{\text{peak requirements per hour}}{\text{batches per hour}}$$

$$= \frac{4\text{m}^3/\text{h}}{24\text{ batches/h}} = 0.17\text{m}^3/\text{h per batch}$$

The concrete mixer capacities are given in litres. Look back to the list of common sizes. *Which one would be needed to keep to programme in this case?*

Now try this example.

Select a suitable concrete mixer for the following conditions: weekly peak requirements 500m³, batching day of $7\frac{1}{2}$ hours, using a pan mixer giving 1 minute mixing time per batch.

TRANSPORTING MACHINES

Once the concrete is mixed it must be taken to where it is needed, and this involves both horizontal and vertical movement. It needs to be taken from the batching plant to the building, lifted from the ground to the floor where it is needed, and placed in its final location. The following forms of transport are widely used: dumper trucks, fork-lift dumper trucks, concrete pumps, cranes and hoists. *Obtain some trade literature on transporting plant. Can you think where each method of transport is likely to be most useful?*

PLACING THE CONCRETE

Vibration is essential in the placing of concrete. *Can you think why?* The main vibrators used are the internal poker vibrator, and the beam vibrator. The picture shows both in use as well as the timber beam tamp. An external vibrator can also be used for pre-cast work.

Once the concrete slab has been vibrated and tamped the surface needs finishing. This is sometimes done by a levelling screed, but savings are made by **direct finishing**.

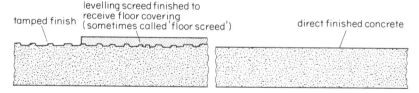

For direct finishing the concrete must be hard enough to take the weight of the finishing machine. However, there needs to be sufficient moisture in the concrete for it to remain workable so that finishing can be achieved. Water can be removed from the concrete at a controlled rate by means of a vacuum mat.

As well as a vacuum mat the illustration shows a power float which is commonly used for direct finishing. A power trowel may also be used to obtain a smooth finish. An alternative would be to allow the concrete to harden fully and then to grind down the surface to a smooth finish using a power grinding machine.

Activity

1 Obtain some trade literature on concrete mixers, transport machines and finishing machines.
2 Select the mixer, type of transporting plant and equipment needed for all concreting operations. List all the equipment needed and locate on the site plan the mixer and static plant. (Use your trade literature if necessary.)

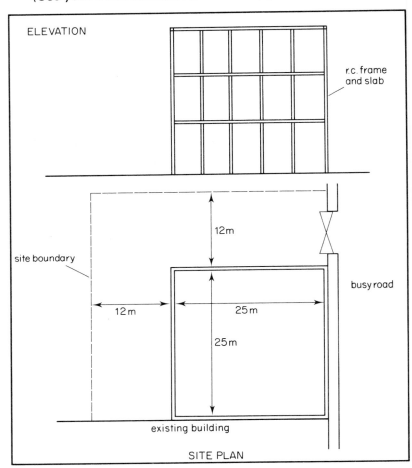

ELEVATION

r.c. frame and slab

site boundary

12m

12m   25m

25m

busy road

existing building

SITE PLAN

STAGE PROGRAMME – FRAME

operation	contract week									remarks
	1	2	3	4	5	6	7	8	9	
excavate	▨									
concrete foundations		▨								30 m³
concrete ground floor			▨							50 m³
columns ground floor				▨						15 m³
shutter & reinforce 1st floor					▨					
concrete 1st floor						▨				125 m³
column 1st floor							▨			15 m³
shutter & reinforce 2nd floor							▨			
concrete 2nd floor								▨		125 m³
columns roof								▨		15 m³
shutter & reinforce roof								▨		
concrete roof									▨	125 m³

10 FORMWORK

THE PURPOSE OF FORMWORK

The purpose of formwork is to contain freshly placed concrete until it has gained sufficient strength to resist superimposed loads, frost damage and mechanical damage. Formwork also helps produce the desired shape and finish to the concrete member.

STRENGTH REQUIRED

Let's look at the first purpose of formwork; to support the fresh concrete.

The formwork must be sufficiently strong to support the loads imposed during placing and curing concrete. These loads will be

1   dead load of the fresh concrete
2   dead load of the formwork.

What other loads will the formwork have to carry?

One load would be the equipment needed for placing the concrete. This could be the plant equipment for transporting the concrete and also the equipment for vibrating the concrete. But that is not all the imposed loading to be carried by the formwork. What effect, for instance, will the vibration have on the formwork? This must be taken into account in calculating the size of the formwork members.

List the four loads to be carried by the formwork.

Producing the desired shape was one of the formwork considerations. At present we are only concerned with formwork to columns, beams and slabs.

The formwork around the beam shape could be detailed so:

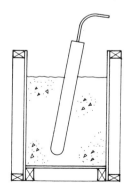

What will happen when the vibrator is started?

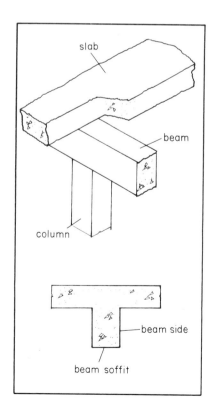

There will be some loss of fine cement slurry between the beam side and soffit formwork. *What effect will this have on the finished concrete member?*

The formwork needs to be a tight fit and to the required tolerances. These tolerances are usually given in the specification for the work.

Economics dictates that the shapes of the members be the same, allowing the formwork to be re-used. The sequence of re-using formwork or moulds is

1 position steel reinforcement *or* position formwork
2 check for alignment plumb and tolerance
3 concrete member
4 cure concrete
5 support until concrete achieves required strength
6 strike formwork
7 clean and re-use.

The time scale will be discussed later but it seems obvious that the more re-uses from the formwork material the cheaper the operation, as long as this is not detrimental to the required finish.

SURFACE FINISH

The finish required to the concrete can affect the cost considerably. Remember the beam section. You were asked to describe the effects on the finished concrete.

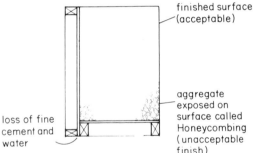

finished surface (acceptable)

aggregate exposed on surface called Honeycombing (unacceptable finish)

loss of fine cement and water

When might this surface defect be acceptable? Different types of sheeting material are also used to create different surface finishes.

Re-using formwork

The type of sheeting material also affects the re-use potential. The Concrete Society has prepared a report dealing with the materials used, the concrete finish and re-use potential.

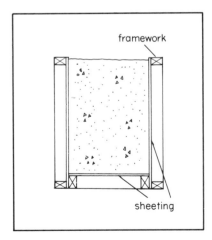

framework

sheeting

Type of plywood	Concrete finish	Re-use potential
Regular	Depends on plywood species but shows grain pattern	up to 10
Coated/painted	High quality exposed surface with uniformity of colour	up to 20
Film-faced	High quality exposed surfaces with uniformity of colour	up to 100

© **Concrete Society. Joint Report on Formwork.**

A case study

A 4 storey office block consists of reinforced concrete fair faced columns. The architect's specification requires a high quality surface to the columns so that decoration can be applied without using plaster. Decide on the type of plywood sheeting you would use and number of re-uses if there are 200 columns, where 10 columns are to be erected every day and it takes a day for the concrete in the forms to set. (Other sheeting materials could be hardboard, timber boards, steel polythene sheeting over plywood and glass reinforced plastic (GRP)).

Did you select the coated/painted plywood? You should have done, since the number of re-uses obtained was ten. Why this answer? Well, how many forms are needed for the operation? Each day ten columns are erected, these must be in position for one day, so the next day another new ten forms to the columns are required. So twenty forms are necessary to keep to required output. The number of re-uses is, therefore, ten.

RELEASING AGENTS

Where the appearance of the concrete is important we must prevent damage to the sheeting when striking or taking down the formwork. To help ease the formwork away from the freshly set concrete we use a **releasing agent**. There are several releasing agents available.

paint sheeting with releasing agent

A neat oil with surfactant is a good general purpose release agent but over application can result in staining of the finished concrete. A surfactant helps increase a liquid's spreading properties. Another release agent is a mould cream emulsion (oil phased, ie oil based on emulsion cream). It is widely used and is recommended for all types of formwork except steel. Chemical release agents like mould cream emulsions are suitable for high-quality finishes. The chemical release agents can be used on steel forms. Another advantage of chemical release agents is the dried coating gives a safer surface to walk on than the oily release agents.

FORMWORK TO COLUMNS

Let's now look at the procedure for formworking, looking at the sheeting first and deciding on the type of plywood, release agent and number of re-uses. For example:

Working out the number of forms

Total number of columns = 500
The works programme gives 50 days for erecting, reinforcing and concreting the columns.

No. of columns per day $= \dfrac{\text{total columns}}{\text{time allowed}} = \dfrac{500}{50} = 10$

The strike time is 1 day, so the number of formworks necessary is as shown overleaf:

Day 1	Day 2	Day 3	Day 4
Erect steel & concrete 10 columns	Cure concrete for 1 day	Strike columns from Day 1. Erect steel & concrete another 10 columns	Cure concrete for 1 day
	Erect steel & concrete 10 columns	Cure concrete for 1 day	Strike column from Day 2. Erect steel & concrete another 10 columns
10 forms required	10 forms required		

So a total of 20 column forms must be made.

$$\text{The number of re-used forms} = \frac{\text{total columns}}{\text{No. of column forms made}} = \frac{500}{20} = 25$$

Using the table on page 72 the formwork sheeting must be film-faced plywood. *What type of releasing agent will be used?*

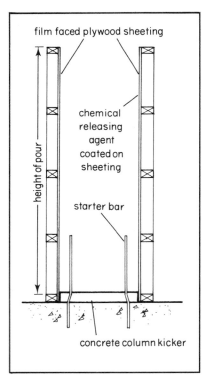

film faced plywood sheeting

chemical releasing agent coated on sheeting

height of pour

starter bar

concrete column kicker

We can now look at the thickness of the plywood and the size of framework required.

The weight of the concrete will depend on the rate of pour. In columns we normally fill the concrete to column full height, whereas in a wall we would fill or pour the concrete in lifts.

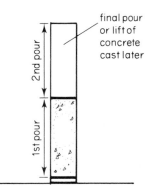

final pour or lift of concrete cast later

2nd pour

1st pour

PLAN OF COLUMN
softwood framing

weight of concrete + vibration load transferred to sheeting

sheeting transfers load to framing

What loads will the sheeting have to resist?

The weight of the wet concrete plus the equipment and vibration load has to be supported.

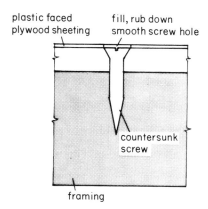

plastic faced plywood sheeting

fill, rub down smooth screw hole

countersunk screw

framing

Fixing the sheeting

How will the framework and the sheeting be fixed? Remember the concrete will faithfully reproduce the sheeting profile, so if you nail the sheeting to the framework what will be shown on the finished surface of the concrete?

Nailing means the formwork can be easily knocked down and remade to different shapes. But in our example the architect wants fair faced concrete. What method will we use for securing the sheeting to the frame? Countersunk and filled screws would be acceptable.

Now we have made the column formwork let's erect the column and concrete it. *List the sequence of operations involved in concreting columns for reinforced concrete framed buildings.*

Positioning the formwork

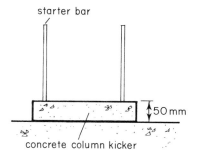

starter bar

50 mm

concrete column kicker

How will we position the formwork for alignment? The starter bars are already through but we cannot use these for alignment. A **concrete kicker** is first cast and then the formwork is erected tight up to the kicker. What is the purpose of the kicker? Next the steel reinforcement to the column is erected.

What sort of typical steel reinforcement would be used?

Before the formwork is erected the release agent is applied either by brush or spray. Why did we choose the use of a chemical release agent rather than a mould cream emulsion or a neat oil with surfactant?

Holding it together

The column formwork needs to be kept tight to prevent shape distortion and loss of cement slurry. A steel **column cramp** can be used.

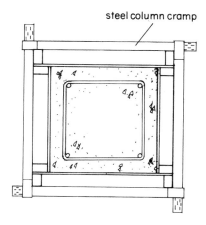

steel column cramp

These will be positioned to take the pressure from the formwork. Where are the maximum and minimum pressures on the formwork? The steel column cramps will be positioned closer together at the base of the column. The column needs to be kept plumb and in alignment. To do this we use steel adjustable props or push pull props.

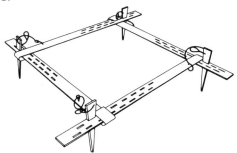

What are labels a *to* g *on the diagram?*

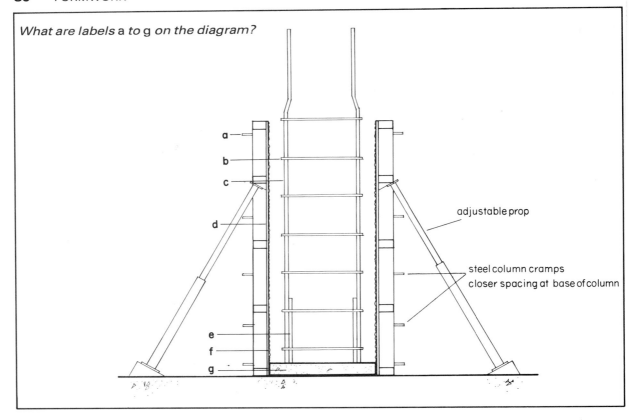

adjustable prop

steel column cramps
closer spacing at base of column

Having erected the formwork we are now ready to concrete the column, or are we? *What will you check prior to concreting?*

The column kicker could have rubbish collecting on it. We don't want this to happen.

Well, how will you clean the top of the kicker?

The column formwork is checked for tightness, alignment and plumb prior to concreting. The steel reinforcement must be covered by a certain thickness of concrete. Can you think why? *How would you ensure this concrete cover?*

We can now pour the concrete.

Review

1. What is the equation for calculating the number of re-uses on formwork?
2. List three types of releasing agents and their disadvantages.
3. List the sequence of operations from setting out to striking column formwork.
4. What is the purpose of a kicker?
5. Sketch a plan and elevation of a column formwork. Label all components.
6. How many re-uses can be expected from these plywood sheets: *a*) regular *b*) coated *c*) film-faced?

CURING CONCRETE

Concrete hardens by 'hydration'. As the water evaporates voids are created in the setting concrete. It is the extent to which these voids are filled with silicate gel that determines the strength, durability and density of the concrete.

As active hydration takes place in the first few hours after placing fresh concrete, it is important for water to be retained during an extended period. This is called **curing**.

The rate of evaporation from an unprotected area will be higher when

a) the relative humidity is low
b) the wind speed is high
c) the concrete temperature is high or not uniform.

Curing is the act of controlling the concrete temperature and water content in the concrete for a definite period of time after placing. The time for curing concrete depends on

a) air temperature
b) shuttering material
c) concrete temperature
d) curing material thermal insulation
e) size of pour.

The optimum concrete temperature is 20°C.

The length of hydration of the cement and, therefore, the rate of hardening of the concrete depends on temperature and moisture available. The duration of controlled curing is important.

Minimum periods of curing OPC (Ordinary Portland Cement)

Weather	Number of days where average temperature of concrete exceeds 10°C
Normal	2 days
Hot or drying winds	4 days
Approaching freezing (below 2°C)	Over 4 days, make sure concrete temperature *at least* 5°C

Plastic cracking is caused by volume changes in the plastic concrete resulting from rapid loss of moisture within a few hours of placing. The top of the concrete member stiffens quicker than the rest, thereby causing tensile stress which the concrete cannot resist, hence plastic cracking.

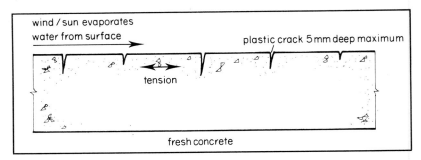

What can be done to reduce plastic cracking?

    a) erect windbreaks
    b) in hot weather, lower the concrete temperature
    c) protect the concrete with wet coverings or impervious sheeting
    d) spray water over concrete.

Review

1   *a*) When the water evaporates what are the voids filled with?
    b) When the voids are left empty what does this affect?
2   Which edge will water evaporate from most quickly?
3   What three factors increase the rate of evaporation from concrete?
4   What is the direction of heat flow from the concrete?

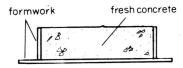

5   Where is curing required and for how long in **a**?
    Where will insulated curing media be required in **b**? What temperature should the concrete be above?

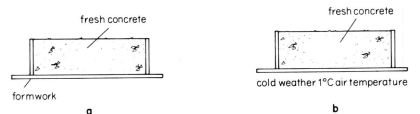

6   What method is needed to reduce plastic cracking of concrete slabs laid in
    a) hot weather, no breeze?
    b) very windy drying weather?

Now use what you have learnt to complete the following case studies.

Case study 1

Weather conditions – Frost is expected during the night, the air temperature was falling during placing of concrete. When last batch was laid air temperature was 3°C.

Site conditions – very exposed site, the poured slab was the roof to a two storey office block.

Guidelines to answer:

1 Location of curing media.
2 Type of material used for curing media, with reasons.
3 Curing period, with reasons.

(See answer on page 211)

Case study 2

Weather conditions – Prolonged, dry, very hot spell. The wet/dry bulb was reading very low relative humidity.

Site conditions – first floor slab of a two storey office block in a restricted city centre site.

Guidelines to answer:

1 Location of curing media.
2 Type of material used for curing media, with reasons.
3 Curing period.
4 Precautions taken during placing of concrete.

(See answer on page 212)

STRIKING FORMWORK

Striking of the formwork is the dismantling of the formwork around the concrete member.

The time which must elapse from pouring the concrete to striking the formwork is called the **strike time**. This time will vary depending on the weather, the exposure of the site, any subsequent treatment to the placed concrete and method of curing. There is another factor which affects the strike time; what is it?

The Code of Practice 110 gives minimum periods before striking formwork.

Formwork	Surface temperature of concrete	
	10°C	7°C
Vertical formwork to columns	9 hours	12 hours
Slab soffits (props left in)	4 days	7 days
Beam soffit (props left in)	8 days	14 days
Props to slab	11 days	14 days
Props to beams	15 days	21 days

Obviously the shorter the strike time the more re-uses of formwork and cost advantages. Research is being done correlating the formwork thermal conductance, the concrete ambient temperature and the type of cement with the concrete strength build-up.

Tables are now available for use on strike times (see *Striking Times for Formwork – Tables of curing periods to achieve given strengths* J Weaver and B M Sadgrove, CIRIA).

Storing formwork

Care must be taken to minimise damage to the sheeting when striking formwork. After the formwork has been stripped it should be cleaned, repaired if necessary and treated with the release agent and stacked.

How will you stack column formwork?

Review

Now complete this exercise.

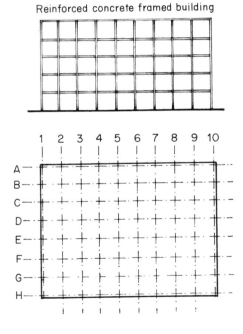

Reinforced concrete framed building

Total columns = 400   Columns per day required = 8

1 What type of plywood sheeting would you use?
2 What is the maximum number of shutter re-use.
3 What type of release agent will be used.
4 Prepare a bar chart to show the sequence of operation and the strike time if temperature was 10°C.

(See answer on page 212)

11 FIRE PROTECTION

THE IMPORTANCE OF FIRE PROTECTION

All structural materials can be damaged in severe fire conditions. Steel and reinforced concrete for instance, although non-combustible and making no contribution to the fire itself, will both be weakened by the heat, and the building may collapse.

The amount of fire protection needed for these materials depends on what the building is used for. The Building Regulations, Part E (Table to Regulation E2) specify different 'purpose groups' for buildings such as

Purpose Group I: Small residential,
Purpose Group II: Institutional.

These purpose groups are used to work out the **fire resistance** which should be given to the structural elements of the building. The fire resistance is the amount of time the building can withstand the heat of the fire before collapsing.

Table to Regulation E2

Designation of purpose groups

Purpose group (1)	Descriptive title (2)	Purposes for which building or compartment is intended to be used (3)
I	Small residential	Private dwelling-house (not including a flat or maisonette)
II	Institutional	Hospital, home, school or other similar establishment used as living accommodation for, or for treatment, care or maintenance of, persons suffering from disabilities due to illness or old age or other physical or mental disability or under the age of five years, where such persons sleep in the premises
III	Other residential	Accommodation for residential purposes other than any premises comprised in groups I and II
IV	Office	Office, or premises used for office purposes, meaning thereby the purposes of administration, clerical work (including writing, book-keeping, sorting papers, filing, typing, duplicating, machine-calculating, drawing and the editorial preparation of matter for publication), handling money and telephone and telegraph operating, or as premises occupied with an office for the purposes of the activities there carried on
V	Shop	Shop, or shop premises, meaning thereby premises not being a shop but used for the carrying on there of retail trade or business (including the sale to members of the public of food or drink for immediate consumption, retail sales by auction, the business of lending books or periodicals for the purpose of gain, and the business of a barber or hairdresser), and premises to which members of the publc are invited to resort for the purpose of delivering there goods for repair or other treatment or of themselves carrying out repairs to, or other treatment of, goods
VI	Factory	Factory within the meaning ascribed to that word by section 175 of the Factories Act 1961(a) (but not including slaughter houses and other premises referred to in paragraphs (d) and (e) of subsection (1) of that section)
VII	Other place of assembly	Place, whether public or private, used for the attendance of persons for or in connection with their social, recreational, educational, business or other activities, and not comprised within groups I to IV
VIII	Storage and general	Place for storage, deposit or parking of goods and materials (including vehicles), and any other premises not comprised in groups I to VII

(a) 1961 c. 34.

Let us calculate the fire resistance for the structural elements of the office building shown.

Look up the purpose group of the office block. Table E5 of the Building Regulations stipulates maximum floor areas, building volume and fire resistance for each purpose group.

What about our office then? Its floor area is 20 m × 10 m = 200 m². Its height is 8 m. Using table E5 we can see that our building does not exceed 15 m in height, that there is no limit on floor area but there is a limit on the cubic capacity of the building, namely, 3500 m³.

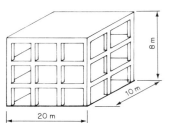

What is the cubic capacity of our building?

$$20 \text{ m} \times 10 \text{ m} \times 8 \text{ m} = 1600 \text{ m}^3.$$

Table to Regulation E5

Minimum periods of fire resistance

In this Table–
CUBIC CAPACITY means the cubic capacity of the building or, if the building is divided into compartments, the compartment of which the element of structure forms part;
FLOOR AREA means the floor area of each storey in the building or, if the building is divided into compartments, of each storey in the compartment of which the element of structure forms part; and
PART, in column (1), means a part which is separated as described in regulation E5(1)(b)

Part 1: Buildings other than single storey buildings

Purpose group		Maximum dimensions			Minimum period of fire resistance (in hours) for elements of structure* forming part of–		
		Height (in m)	Floor area (in m²)	Cubic capacity upper storey	ground storey or	basement storey	
(1)		(2)	(3)	(4)	(5)	(6)	
I	Small residential:						
	House having not more than three storeys	No limit	No limit	No limit	½	1†	x
	House having four storeys	No limit	250	No limit	1‡	1	
	House having any number of storeys	No limit	No limit	No limit	1	1½	
II	Institutional	28	2,000	No limit	1	1½	
		over 28	2,000	No limit	1½	2	
III	Other residential:						
	Building or part having not more than two storeys	No limit	500	No limit	½	1	x
	Building or part having three storeys	No limit	250	No limit	1‡	1	
	Building having any number of storeys	28	3,000	8,500	1	1½	
	Building having any number of storeys	No limit	2,000	5,500	1½	2	
IV	Office	7·5	250	No limit	½	1†	x
		7·5	500	No limit	½	1	
		15	No limit	3,500	1‡	1	
		28	5,000	14,000	1	1½	
		No limit	No limit	No limit	1½	2	
V	Shop	7·5	150	No limit	½	1†	x
		7·5	500	No limit	½	1	
		15	No limit	3,500	1‡	1	
		28	1,000	7,000	1	2	
		No limit	2,000	7,000	2	4	y
VI	Factory	7·5	250	No limit	½	1†	x
		7·5	No limit	1,700	½	1	
		15	No limit	4,250	1‡	1	
		28	No limit	8,500	1	2	
		28	No limit	28,000	2	4	
		over 28	2,000	5,500	2	4	

For the purpose of regulation E5(2), the period of fire resistance to be taken as being relevant to an element of structure is the period included in column (5) or (6), whichever is appropriate, in the line of entries which specifies dimensions with all of which there is conformity or, if there are two or more such lines, in the topmost of those lines.

* A floor which is immediately over a basement storey shall be deemed to be an element of structure forming part of a basement storey.

† The period is half an hour for elements forming part of a basement storey which has an area not exceeding 50 m².

‡ This period is reduced to half an hour in respect of a floor which is not a compartment floor, except as to the beams which support the floor or any part of the floor which contributes to the structural support of the building as a whole.

x The line of entries thus marked is applicable only to buildings, not to compartments, except in relation to purpose group III; see also regulations E7(3) proviso (i) and E8(7) proviso (a).

y If the building is fitted throughout with an automatic sprinkler system which complies with the relevant recommendations of CP 402.201: 1952, any maximum limits specified in columns (3) and (4) shall be doubled.

Table to Regulation E5 – continued

Minimum periods of fire resistance

Part 2: Single storey buildings

Purpose group (1)		Maximum floor area (in m²) (2)	Minimum period of fire resistance (in hours) for elements of structure (3)	
I	Small residential	No limit	$\frac{1}{2}$	z
II	Institutional	3,000	$\frac{1}{2}$	z
III	Other residential	3,000	$\frac{1}{2}$	z
IV	Office	3,000 No limit	$\frac{1}{2}$ 1	z
V	Shop	2,000 3,000 No limit	$\frac{1}{2}$ 1 2	z
VI	Factory	2,000 3,000 No limit	$\frac{1}{2}$ 1 2	z
VII	Assembly	3,000 No limit	$\frac{1}{2}$ 1	z
VIII	Storage and general	500 1,000 3,000 No limit	$\frac{1}{2}$ 1 2 4	z

For the purpose of regulation E5(2), the period of fire resistance to be taken as being relevant to an element of structure is the period included in column (3) in the line of entries which specifies the floor area with which there is conformity or, if there are two or more such lines, in the topmost of those lines.

z See regulations E7(3) proviso (i) and E8(7) proviso (b).

It is clear we are within the requirements of the building regulations; that is, if we construct our building with minimum fire resistance periods of one hour for both the ground floor and the basement.

Using the Building Regulations calculate the fire resistance required for this building.

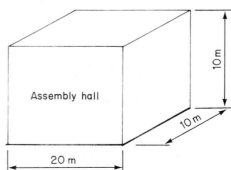

Assembly hall

10 m

10 m

20 m

Having obtained the required fire resistance to be given to the structural elements, we must now consider methods of construction and materials which will give the structural frame the required fire resistance.

FIRE PROTECTION MATERIAL

Traditionally structural steel was encased in concrete. The concrete encasing did not take any of the building load, it merely served to give the structural steel the necessary fire resistance.

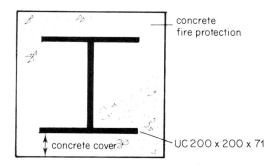

concrete fire protection

concrete cover

UC 200 x 200 x 71

SOLID PROTECTION

The more **cover** or thickness of concrete to the universal column the greater the fire resistance given. Building Regulations stipulate the thickness or cover to be given using concrete as a fire protective casing. This is called **solid** protection as opposed to **hollow** fire protection. Now look at the Building Regulations, Schedule 8, Part V: Structural steel, on page 90. This part of Schedule 8 deals with solid protection. Firstly the mix of concrete to be used is specified. See table (A) solid protection (unplastered).

What is the minimum mix of concrete to be used as solid protection?

The table states that the concrete encasing can be either non-loadbearing or loadbearing. Traditionally the concrete was non-loadbearing, and this is now considered a disadvantage of solid fire protection. Both the load bearing and the non-loadbearing concrete encasing has to be reinforced.

What type of reinforcement must be included with solid concrete fire protection?

The required fire resistance for the office building we are considering was 1 hr. Using the Schedule 8 of the Building Regulations the cover or thickness of concrete which is non-loadbearing must be 25 mm thick for the 1 hour fire resistance.

What mix of concrete and weight of wire mesh should be included in the solid concrete fire protection? From the Building Regulations Schedule 8, Part V: Structural steel, note the other types of solid protection available.

Schedule 8 Part V: Structural steel

A. Encased steel stanchions (mass per metre not less than 45 kg)
(Note: In the following table, figures in brackets are applicable only in relation to universal columns of serial size 203×203 (8×8) as designated in BS 4: Part 1: 1972)

Construction and materials	Minimum thickness (in mm) of protection for a fire resistance of–				
	4 hours	2 hours	1½ hours	1 hour	½ hour
(A) Solid protection* (unplastered)					
1. Concrete not leaner than 1:2:4 mix with natural aggregates–					
(a) concrete not assumed to be loadbearing, reinforced†	50	25	25	25	25
(b) concrete assumed to be loadbearing, reinforced in accordance with BS 449: Part 2: 1969 ¶	75	50	50	50	50
2. Solid bricks of clay, composition or sand-lime	75	50	50	50	50
3. Solid blocks of foamed slag or pumice concrete reinforced† in every horizontal joint	62	50	50	50	50
4. Sprayed asbestos of density 140–240 kg/m³	(70)	(30)	(25)	(20)	(10)
5. Sprayed vermiculite-cement		38	32	19	12·5
(B) Hollow protection‡					
1. Solid bricks of clay, composition or sand-lime reinforced in every horizontal joint, unplastered	115	50	50	50	50
2. Solid blocks of foamed slag or pumice concrete reinforced § in every horizontal joint, unplastered	75	50	50	50	50
3. Metal lathing with gypsum or cement-lime plaster of thickness of		38§	25	19	12·5
4. (a) Metal lathing with vermiculite-gypsum or perlite-gypsum plaster of thickness of	50§	19	16	12·5	12·5
(b) Metal lathing spaced 25 mm from flanges with vermiculite-gypsum or perlite-gypsum plaster of thickness of	44	19	12·5	12·5	12·5
5. Gypsum plasterboard with 1·6 mm wire binding at 100 mm pitch–					
(a) 9·5 mm plasterboard with gypsum plaster of thickness of				12·5	12·5
(b) 19 mm plasterboard with gypsum plaster of thickness of		12·5	10	7	7
6. Gypsum plasterboard with 1·6 mm wire binding at 100 mm pitch–					
(a) 9·5 mm plasterboard with vermiculite-gypsum plaster of thickness of		16	12·5	10	7
(b) 19 mm plasterboard with vermiculite-gypsum plaster of thickness of	32§	10	10	7	7
7. Metal lathing with sprayed asbestos of thickness of	(70)	(30)	(25)	(20)	(10)
8. Vermiculite-cement slabs of 4:1 mix reinforced with wire mesh and finished with plaster skim. Slabs of thickness of	63	25	25	25	25
9. Asbestos insulating boards of density 510–880 kg/m³ (screwed to 25 mm thick asbestos battens for ½ hour and 1 hour periods)		25	19	12	9

* Solid protection means a casing which is bedded close to steel without intervening cavities and with all joints in that casing made full and solid.
† Reinforcement shall consist of steel binding wire not less than 2.3 mm in thickness, or a steel mesh weighing not less than 0.48 kg/m². In concrete protection, the spacing of that reinforcement shall not exceed 150 mm in any direction.
‡ HOLLOW PROTECTION means that there is a void between the protective material and the steel. All hollow protection to columns shall be effectively sealed at each floor level.
§ Light mesh reinforcement required 12.5 mm to 19 mm below surface unless special corner beads are used.
¶ As read with Addendum No.1 (April 1975) to BS 449: Part 2: 1969 and Supplement No.1 (PD 3343) to BS3449: Part 1: 1970.

LIGHTWEIGHT FIRE PROTECTION

The use of solid protection such as concrete encasing of structural steel work has some disadvantages. Can you think of one disadvantage of using concrete solid fire protection which affects the cost of foundations, hence building costs?

The use of lightweight fire protection avoids this disadvantage of solid protection – additional dead load. This type of fire protection is described as hollow protection.

hollow fire protection

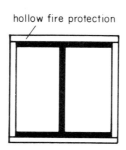

solid fire protection

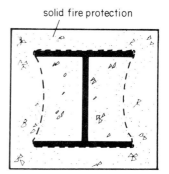

What are some of the advantages claimed by the manufacturers of hollow lightweight fire protection? It is less bulky usually than traditional methods of encasing. The materials are lightweight, thereby saving cost incurred in foundations. Some are pre-finished and do not involve wet finishings, allowing decoration to commence immediately. Finally, some manufacturers claim speed of erection and consequent saving in site costs as two important advantages of lightweight hollow protection. The Building Regulations can be used as a guide for hollow fire protection. Schedule 8, Part V: Structural steel, table (B), in the 1976 Building Regulations recommends materials for hollow protection. Some of the materials do not fall into the category of lightweight hollow protection.

Using Schedule 8, Part V, list six lightweight hollow protections recommended.

An alternative to using the Building Regulations is to use manufacturer's literature. Still the Building Regulations must always be complied with and therefore used to obtain the fire resistance for specific structural elements.

Review

1 Why do non-combustiblbe structural materials such as steel and concrete need fire protection?
2 What is the sequence used to obtain the fire resistance of a building using Building Regulations?
3 Name the two methods of protection available to give the required fire resistance to structural elements.

12 PANEL INFILL

WHAT IS PANEL INFILL?

Panel infill is a term which applies only to a framed building. The wall enclosure function is performed by a panel of suitable material and construction, infilling between the columns and beams. The structural frame is exposed on the elevation as well as the wall infill now called panel infill. The completed elevation will look like this.

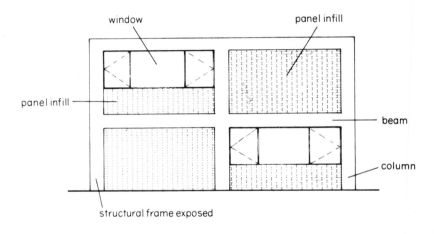

ITS FUNCTIONS   *What are the performance requirements for the panel infill?*

The panel infill will have to perform several functions. Imagine you are the man working in the office above. *List three functions or requirements you would like the panel infill to perform.*

The wall function depends on the use of the building and its structure. Some performance requirements will have priority over other requirements and thus affect the choice of material used. In this section we are only concerned with brick panel infill.

Here are some wall performance requirements:

1	strength and stability	5	thermal insulation
2	durability	6	sound insulation
3	dimensional stability	7	fire resistance
4	exclusion of water	8	appearance.

In a framed building the wall does not take any structural load – the frame takes the dead and live loads imposed on the building. *Why consider the wall performance requirements of strength and stability?*

The photograph will give the reason for considering strength and stability. The damage was caused by the wind. The wind can also suck out a wall. The wind pressure may be either negative (suction) or positive pressure. The site location will affect the wind pressure. It is usual to consider wind loading in conjunction with Code of Practice 3.

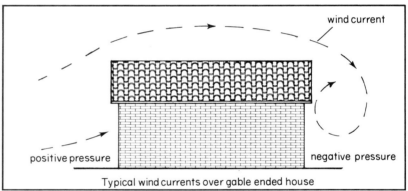

Typical wind currents over gable ended house

CONNECTING THE FRAME TO THE PANEL

The structural frame has been designed to carry the windloading to the foundation. The windload on the panel infill is transmitted to the frame using anchors.

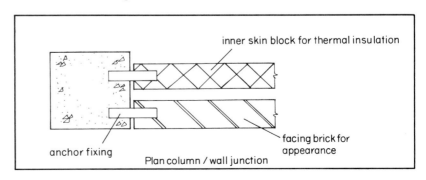

inner skin block for thermal insulation

anchor fixing

facing brick for appearance

Plan column / wall junction

A typical anchor fixing is the dovetail slot and anchor.

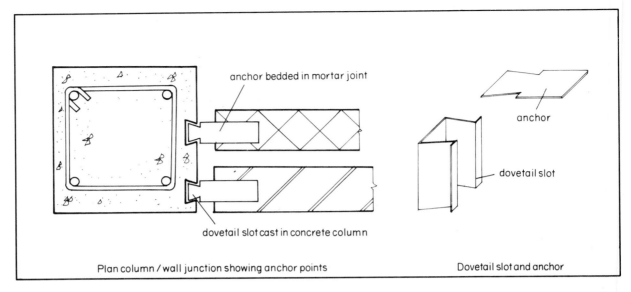

Plan column / wall junction showing anchor points

Dovetail slot and anchor

The fixing depends on the loadings anticipated. A suggestion would be to have a dovetail slot and anchor fixing every third course of brickwork.

Why cannot the brickwork and the concrete column butt up together? One reason is due to component tolerances. The brick length according to British Standard can vary in length. The column is not true to its design dimension – there is a tolerance.

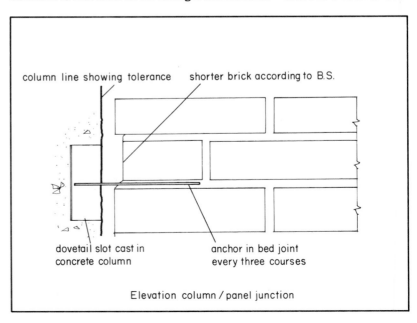

Elevation column / panel junction

Another reason is the differential expansion and contraction of these two building components. The structural frame will move, as will the panel infill material. A joint is specially incorporated to accommodate movement and is known as an expansion or movement joint.

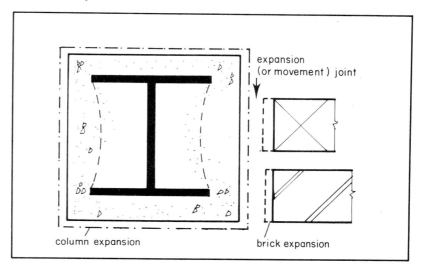

column expansion   brick expansion

SEALANTS It is common practice to use mortar as a jointing material between the structural frame and the panel infill. Another alternative is to use a **mastic sealant** i.e. a sealant which has a degree of flexibility. There are a number of ways of classifying sealants. The usual method is by the basic chemical ingredient, although classification can be by performance properties.

Sealants can be divided into two broad categories:

    1   plastic   2   elastomeric.

The plastic materials are commonly referred to as mastics, have no recovery properties and tend to be the lower performance materials. Elastomeric sealants cure to a rubbery set and show some recovery and are higher performance sealants. It is this material, elastomeric, which we will be considering.

Polysulphide rubber sealants

This sealant can be two part or one part. The two part pack (or multi-part pack) is where the curing agent is mixed into the base immediately prior to application. There is a curing agent pack and a base pack, hence the term two part sealant.

Polysulphide rubber sealants are used for expansion joints, construction joints and window pointing, particularly in situations where maintenance may be difficult, or severe exposure is expected. The joint is designed with depth equal to half the width.

Minimum joint size is 6 mm×6 mm deep, the maximum probably 25 mm×12 mm deep.

Acrylic resin sealants

These acrylic resin sealants have similar properties to the polysulphide rubber sealants, if high performance acrylics are used. The joint is designed so that the depth is equal to half the width with a minimum of 6 mm×6 mm.

Silicone sealants

Silicone sealant has the same application as acrylic resin or polysulphide sealants but may be more expensive.

Polybutadiene sealant

These sealants are usually two part sealants. It is claimed that polybutadiene sealants have the same applications as polysulphide sealants.

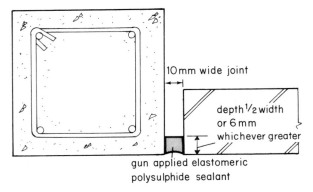

10 mm wide joint

depth ½ width or 6 mm whichever greater

gun applied elastomeric polysulphide sealant

Plan of column/panel junction showing position of sealant

The joint needs a back up material. If, for example, the joint was backed up with mortar the flexibility of the joint would be impaired. A back up material such as bitumen impregnated fibre board would allow movement of the panel infill and the structural frame.

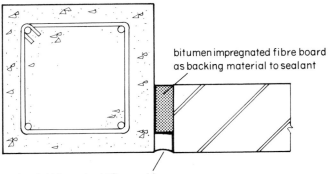

bitumen impregnated fibre board as backing material to sealant

polysulphide sealant 10 mm wide 6 mm deep gun applied sealant

Class of sealant	Expected life years	Uses	Joint size mm	Adhesion	Installation
Polysulphide	25+	Expansion construction joints, pointing.	6×6 min 25 max	Good. Excellent with primer.	Fair. Gun applied.
Silicone	20	Pointing	6×6 min 25 max	Excellent when used with correct primer.	Good. Gun applied.
Polybutadiene	20	Pointing	6×6 min 25 max	Good. Primers generally required.	Fair. Gun or knife applied.

We have left out important considerations such as life expectancy, adhesion of the sealant to substrata, and installation.

What about the other performance requirements? The remaining performance requirements such as durability, exclusion of water, thermal insulation and appearance, can be determined by the choice of material and the construction of the wall.

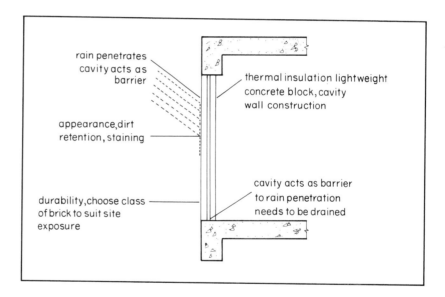

In considering rain penetration and durability the site is classified by the degree of exposure. The classes of exposure are

    1 sheltered   2 moderate   3 severe.

The site exposure will affect the class of brick selected, along with the rain index (that is, the amount of rainfall and its orientation) and the type of construction.

WHAT CLASS OF BRICK? Read through carefully the Building Research Establishment Digests 164, *(Clay brickwork: 1)* and 165, *(Clay brickwork: 2)* starting overleaf.

Building Research Establishment Digest 164

Clay brickwork: 1

Varieties, qualities and types

Clay bricks are classified in BS 3921: Part 2: 1969 *Bricks and blocks of fired earth, clay or shale* as follows:

Varieties: *common* – suitable for general building work

facing – made or selected for appearance

engineering – a dense, strong brick conforming to defined limits of absorption and strength

Qualities: *internal* – suitable only for internal use

ordinary – normally durable in the external face of a building

special – durable in conditions of extreme exposure to water and freezing

Types: *solid* – small* holes not exceeding 25 per cent of the volume of the brick are permitted; alternatively, frogs not exceeding 20 per cent of the total volume are permitted

perforated – small* holes may exceed 25 per cent of the total volume of the brick

hollow – the total of holes, which need not be small, may exceed 25 per cent of the volume of the brick

cellular – holes closed at one end exceed 20 per cent of the volume

The widening of the definition 'solid' to include bricks with frogs or small holes is not consistent in all sources of information. For example, BS 743: 1970 *Materials for damp-courses* calls for bricks with no holes, as does Table 5 of CP 121: Part 1: 1973 *Notional fire resistance of walls*, although Building Regulations 1972, in Schedule 8 *Notional periods of fire resistance* refers to solid bricks but allows 25 per cent perforation. Furthermore, three hand-holes, each up to 3250 mm² cross-section (equivalent to 64 mm diameter), are allowed by the Standard within the 25 per cent and these are over half the width of the brick.

'Special' refers to a defined quality of brick, although bricks of shapes other than rectangular prisms are referred to as 'standard specials' and BS 4729: 1971 gives dimensions for a range of these special

*'small' holes are less than 20 mm wide or less than 500 mm² in cross-section

shaped bricks in relation to the Imperial size of brick given in BS 3921 Part 1 and the co-ordinated metric size of brick given in BS 3921: Part 2. Note that the two sets of dimensions, although very similar, are not straight Imperial to metric conversions and the bricks are not truly interchangeable.

Sizes

Work sizes and co-ordinating sizes of bricks to BS 3921 are illustrated in Fig 1.

The tolerances on the sizes of bricks are fixed by giving maximum and minimum dimensions, not on individual bricks, but on batches of 24 bricks chosen at random, as in Table 1.

Fig 1 Work sizes (firm lines) and co-ordinating sizes (broken lines) of bricks to BS 3921

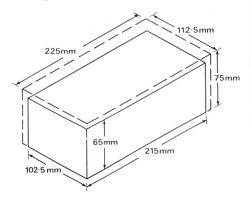

It follows statistically from this method of measurement that batches are unlikely to contain more than 1 per cent of bricks outside the limit for length of ±6 to 7 mm. Nevertheless, for critical work the bricklayer may need to recognise a much longer or shorter brick if it occurs and not to use it, thus avoiding complaints about the variation of perpends.

The brick sizes in BS 3921 are as yet only co-ordinated metric sizes. Proposals for the designation and work sizes of metric modular bricks have been made in the BS Draft for development 'Clay Bricks with Modular Dimensions' and are quoted in Table 2. Some or all of these brick sizes may already be available in some areas.

The proposed limits on the dimensions of these bricks, measured over a random sample of 24 bricks, are given in Table 3.

Weight

Clay brickwork weighs from about 2 tonnes/m³ (2 kg/m² per mm thickness) for typical common bricks to 2·4 tonnes/m³ for bricks of high density (eg engineering bricks).

Table 1 Overall measurement of 24 bricks

	maximum	minimum
	mm	mm
length	5,235	5,085
width	2,505	2,415
height	1,620	1,530

Table 2 Proposed sizes for metric modular bricks

	work size		
Designation	length	width	height
	mm	mm	mm	mm
300×100×100	288	90	90
200×100×100	190	90	90
300×100×75	288	90	65
200×100×75	190	90	65

Table 3 Proposed size limits for metric modular bricks

		limits of size (24 bricks)	
Work sizes		maximum	minimum
	mm	mm	mm
length	288	7,012	6,812
	190	4,626	4,494
width	90	2,205	2,115
height	90	2,205	2,115
	65	1,605	1,515

Appearance

Where colour and texture are important, these should be agreed on the basis of samples representative of the production range. Occasionally there may occur bricks of attractive colour such as salmon pink which might indicate comparative underfiring of the bricks. Such bricks may lack durability and, whilst a few of these may be acceptable in work subjected to normal exposure, it might prove unwise to use bricks all of this kind, selected on grounds of colour, in conditions of more severe exposure.

Ordinary quality bricks are required by BS 3921 to be reasonably free from deep or extensive cracks, from damage to edges and corners and from pebbles. A more exacting standard is set for special quality bricks. For many purposes, slight cracking will have little effect on strength or resistance to rain and the main criterion is aesthetic but in highly stressed brickwork only a very minor amount of cracks should be allowed.

The standard also calls for the bricks to be free

from expansive particles of lime. This is to avoid the cracking which might occur when the lime hydrates after the bricks have been built in. Experience has shown, however, that when lime particles smaller than 3 mm diameter hydrate they produce only a small 'pock mark' which, provided that there are not many of them, can usually be ignored. Particles larger than this might, if present in any quantity, cause unsightly blemishes or even severe cracking.

A cut (not broken) surface of a brick should show a reasonably uniform texture but this does not imply that the particles should be of uniform size.

The standard for internal quality bricks is less severe and they may not be suitable for fair-faced work; if required for this purpose, this should be specified.

Quality of bricks for elements of brickwork

Table 4 shows the minimum qualities of bricks and the recommended mortars for durability, whilst Table 5 gives the proportions by volume of the mixes recommended in Table 4. These recommendations are illustrated in Fig 2.

Characteristics

Compressive strength

BS 3921 specifies a minimum strength of 5.2 MN/m^2 for bricks; this is sufficient for the loadings in low-rise housing and similar buildings and ensures that the bricks are strong enough to be handled. Higher strength should be specified only when it is going to be used. Table 6 classifies bricks in terms of strength. At the lower end of the range, where the strengths are only about 7 MN/m^2 apart, normal variations will give a wide overlap of strength of individual specimens from different classes. Strength is not necessarily an index of durability. Bricks of Class 7 and upwards are usually durable, but there are bricks approaching this strength which decay rapidly if exposed to freezing in wet conditions and others, very much weaker, which are durable.

Fire resistance

Schedule 8 of the Building Regulations 1972 groups together clay, concrete and calcium silicate brickwork as having the same order of fire resistance, but there are some differences between Schedule 8 and CP 121: Part 1: 1973 for the minimum thickness of unplastered and plastered brick walls. Thus, Schedule 8 provisions will meet the mandatory requirements but CP 121 is also useful because it recommends thicknesses of walls built with bricks in which the perforations exceed the 25 per cent volume limit assumed in Schedule 8.

Table 4 Minimum qualities of bricks and recommended mortar groups (i–v of Table 5)

Constructional element	Early frost hazard[a] no brick	mortar	yes brick	mortar
Internal walls and inner leaf of cavity walls	internal	v	ordinary	iii or plasticised iv
Backing to external solid walls	internal	iv	ordinary	iii or plasticised iv
External walls; outer leaf of cavity walls:				
– above damp-proof course	ordinary	iv[c]	ordinary	iii[c]
– below damp-proof course	ordinary	iii[d, g]	ordinary	iii[b, d, g]
Parapet walls; free-standing walls; domestic chimneys:				
– rendered[e]	ordinary	iii[f]	ordinary	iii[f]
– not rendered	special	ii	special	ii
	preferred or ordinary	iii		
Sills and copings; earth-retaining walls backfilled with free-draining material	special	i	special	i

Notes

(a) During construction, before mortar has hardened (say 7 days after laying) or before the wall is protected against the entry of rain at the top.

(b) If the bricks are to be laid wet, *see* 'Cold weather bricklaying', Digest 160.

(c) If to be rendered, lay in mortar not weaker than group iii, preferably with sulphate-resisting cement.

(d) If sulphates are present in the groundwater and ordinary quality bricks are used, use sulphate-resisting cement in the mortar.

(e) Parapet walls of clay bricks should not be rendered on both sides; if this is unavoidable, select mortar as though *not* rendered.

(f) If the presence of sulphates in the bricks is suspected, group iii mortar made with sulphate-resisting cement is preferred.

(g) CP 121 considers the zone of brickwork more than 150 mm above ground level and below damp-proof course to be at greater risk and suggests the use of 'special' quality bricks.

Table 5 Mortar mixes (proportions by volume)

	Mortar group	Cement : lime : sand *	Masonry-cement : sand	Cement : sand, with plasticiser
Increasing strength but decreasing ability to accommodate movements caused by settlement, shrinkage, etc	i	1:0–$\frac{1}{4}$:3	—	—
	ii	1:$\frac{1}{2}$:4–4$\frac{1}{2}$	1:2$\frac{1}{2}$–3$\frac{1}{2}$	1:3–4
	iii	1:1:5–6	1:4–5	1:5–6
	iv	1:2:8–9	1:5$\frac{1}{2}$–6$\frac{1}{2}$	1:7–8
	v	1:3:10–12	1:6$\frac{1}{2}$–7	1:8

equivalent strengths within each group

increasing frost resistance

improving bond and resistance to rain penetration

Direction of changes in properties

Where a range of sand contents is given, the larger quantity should be used for sand that is well graded and the smaller for coarse or uniformly fine sand.

Because damp sands bulk, the volume of damp sand used may need to be increased. For cement: lime: sand mixes, the error due to bulking is reduced if the mortar is prepared from lime: sand coarse stuff and cement in appropriate proportions; in these mixes 'lime' refers to non-hydraulic or semi-hydraulic lime and the proportions given are for lime putty. If hydrated lime is batched dry, the volume may be increased by up to 50 per cent to get adequate workability.

*The addition of an air-entraining agent might improve the frost-resistance of cement: lime: sand mortars.

Fig 2 Minimum qualities of clay brick for various positions

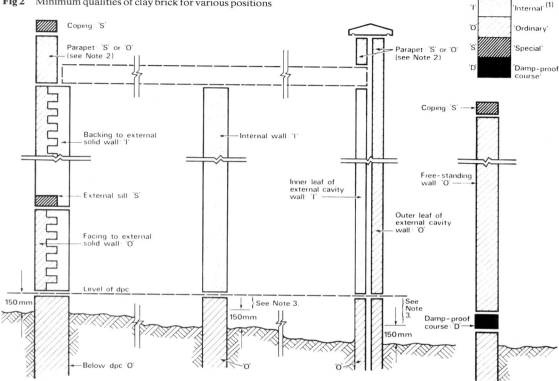

Notes

1 If there is early frost risk, use 'ordinary' quality
2 Parapets: rendered on one side only, 'ordinary' not rendered, 'special' preferred and necessary with early frost risk

rendered on both sides, not desirable but if unavoidable use 'special'
3 CP 121 considers this zone to be at greater risk and suggests the use of 'special' quality bricks

Absorption

Water absorption does not necessarily indicate the behaviour of a brick in weathering. There are no specific requirements in BS 3921 for the absorption of clay bricks other than engineering bricks (*see* Table 6) and bricks for use as damp-proof courses. Low absorption, ie less than 7 per cent by weight, usually indicates a high resistance to damage by freezing, although some types of bricks of much higher absorption may also be frost resistant. No simple correlation can be offered – with some types of bricks a range of absorption from 8–12 per cent by weight will cover the range from resistant to non-resistant to frost damage; on the other hand another type of brick with a range of absorptions from 12–18 per cent might still be frost-resistant even at the higher absorption.

Table 6 Strength and absorption

Designation	Class	Average compressive strength MN/m² not less than	Average absorption (boiling or vacuum) per cent weight, not greater than
Engineering brick	A	69·0	4·5
	B	48·5	7·0
Load-bearing brick	15	103·5	No specific requirements
	10	69·0	
	7	48·5	
	5	34·5	
	4	27·5	
	3	20·5	
	2	14·0	
	1	7·0	

Building Research Establishment Digest 165

Clay brickwork: 2

Characteristics (continued)

Thermal and moisture movements

Thermal expansion The coefficient of linear thermal expansion of clay brickwork has been taken as 5×10^{-6} per °C; it is, however, difficult to predict accurately the likely movement of a wall because of the restraint due to internal friction or internal restriction. Further, because the two surfaces of a wall are often heated unequally, there may be a tendency to differential movement within the wall thickness. Although thermal movement is, in theory, reversible, this may not be wholly true of brickwork in practice because of friction and other effects.

Moisture expansion As clay bricks cool after firing in the kiln they start to take up atmospheric moisture and expand, at a high rate initially but thereafter decreasing, though the movement may continue at a slow and diminishing rate indefinitely. Measurements made at BRS have shown that a typical brick would be expected to expand by about 0·8 mm per metre in the first eight years, of which about half occurs in the first week. There is considerable variation in behaviour between bricks of different origins; thus, some engineering bricks, if only moderately fired, can have larger expansions of up to, say, 1·6 mm per metre, though if well fired to a low absorption the same bricks will usually give very low expansions.

Recent work suggests that the ratio of brickwork expansion to brick expansion is about 0·6, provided no other source of expansion (such as sulphate expansion of the mortar) is also present, and this is considered later under 'Design'. Superimposed on this moisture movement is a small reversible wetting and drying movement which is not likely to exceed 0·02 per cent. There is no evidence that this reversible expansion lessens with time.

The long-term moisture expansion of bricks and brickwork is not accelerated by dipping the bricks in water and is not reversed when they dry out.

Efflorescence

Very small amounts of salts, usually sulphates, which may be present in the bricks and alkalis from the cement used in the mortar are sufficient to produce an efflorescence during the period when a building is drying out. This is likely to be unsightly rather than harmful and should eventually disappear although sulphate efflorescences may recur in spring and autumn for some years.

Efflorescence is destructive only in exceptional instances where the soluble salts crystallise just below the brick surface which, if weak, may then crumble.

If bricks become saturated before the work is completed, the probability of subsequent efflorescence is increased. Brick stacks should, therefore, be protected from rain at all times; during laying, the bricks should be moistened only to the extent that is found absolutely essential to obtain adequate bond between bricks and mortar; newly built work should be protected from rain. The risk of efflorescence may be reduced by using bricks of low soluble salts content but a more useful indication of behaviour can be obtained by making the efflorescence test described in BS 3921.

The most common places for efflorescence to occur are where brickwork may get wet, for example, parapets and earth-retaining walls, and where deficiencies in design may allow excessive water to get into the brickwork.

Efflorescence should always be dry brushed away before rendering or plastering a wall; wetting it will carry the salts back into the wall to reappear later.

Staining

Some bricks may produce rust stains after building in; the cause is the presence of ferrous salts, often but not necessarily associated with dark-coring. The staining may show on the bricks but is more noticeable on light-coloured mortar joints. The risk of staining is accentuated by saturation during storage or during construction. If staining is likely from the reputation of the bricks, the best safeguard is to keep the bricks as dry as possible. In critical work, it might pay to leave the joints raked back for 6–8 weeks after construction to allow the stain to develop on the raked-back mortar, and subsequently point, wetting the joints as little as possible. With time, the solubility of the iron in the bricks diminishes and further risk of staining is negligible.

Sulphate attack

Sulphate attack on brickwork is the result of the reaction of tricalcium aluminate, present in all ordinary Portland cements, with sulphates in solution. Its effect is an overall expansion of the brickwork due to expansion of the mortar joints, followed in more extreme cases by progressive disintegration of the mortar joints. A vertical expansion of up to 0·2 per cent may occur in facing brickwork and as much as 2 per cent in rendered work; horizontal expansion is rather less but may be more obvious. Serious attack is rarely noticeable in less than two years and can thus usually be distinguished from moisture expansion which becomes apparent, if at all, in the first few months after building.

BS 3921 limits the amount of soluble sulphates only in 'special' quality bricks and it must be assumed, unless the manufacturer can produce reliable evidence to the contrary, that all other quality bricks contribute to sulphate attack.

The effect of sulphate attack on mortars can be minimised by specifying the richer mixes, for example, 1:0–¼:3 or 1:½:4–4½, or (better still) 1:5–6 with plasticiser instead of lime, and by using sulphate-resisting Portland cement, which is low in tricalcium aluminate, supersulphated or, in certain instances, high alumina* cement.

Construction water is not normally sufficient to cause sulphate attack but repeated wetting and drying over a period of years is closely correlated with it. Parapets and free-standing walls are most likely to be affected but in regions exposed to driving rain (*see* Digest 127) all external brickwork is potentially at risk if the other necessary factors are present.

Sulphate attack on brickwork is dealt with in more detail in Digest 89.

Frost resistance

BS 3921 gives only partial guidance on the behaviour of bricks in freezing conditions. Bricks for internal walls are not required to resist frost and may, therefore, need protection if stacked on site during the winter. Bricks of ordinary quality are suitable for exposure in external walls between roof and ground-level damp-proof course and should resist frost for one winter whilst stacked on a building site; it is assumed that in service they are unlikely to be wet enough to be damaged by freezing. Bricks of special quality are suitable for situations where they may become and remain very wet, and may be frozen in that condition.

There is no frost test in BS 3921 but bricks of special quality are deemed to be frost-resistant if they satisfy one of the following requirements:

(*a*) The manufacturer shall provide evidence that bricks of the quality offered have given satisfactory service in conditions at least as severe as those proposed for not less than three years in the locality for which their use is being considered.

(*b*) In the absence of such evidence, sample panels should be built in an exposed position under

*Lime must not be added to high alumina cement.

independent supervision. Bricks that behave satisfactorily for not less than three years can be regarded as frost-resistant.

(*c*) Where neither of the foregoing is possible, a brick that is of engineering classification either as regards strength or water absorption shall be deemed to be frost-resistant.

Over most of the British Isles, frost damage is rare in external walls between damp-proof course and eaves, though it is not uncommon in parapets, free-standing walls and retaining walls where care has not been taken in selecting the bricks. In areas of severe exposure, frost damage can occur between damp-proof course and eaves, and even in window reveals; local experience is often the best guide to the suitability of particular bricks in such areas.

Testing and control

Soluble salts content   If a single characteristic such as the soluble content of all the bricks in a load were to be determined, the results would give a distribution curve of the form shown in Fig 3; but testing on this scale is clearly impracticable. The result for a single brick might lie anywhere within this range and would therefore be meaningless but BS 3921 reduces to an acceptable level the chances of unrepresentative sampling by requiring the sample for soluble salts analysis to be prepared from ten bricks, representing a delivery of 2000 to 10 000 bricks.

Fig 3   Normal distribution curve

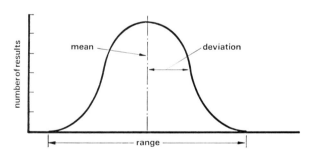

Strength   In the same way, ten bricks are taken for the compressive strength test. Although it may be found that an individual brick varies by 20 per cent or more from the average, the permissible stresses allowed by CP 111: Part 2: 1970 *Structural recommendations for loadbearing walls* take account of this, being based on the average strength of ten bricks. It is, therefore, both unnecessary and uneconomic to insist that every brick is above a certain strength. The

Fig 4 Control charts for strength

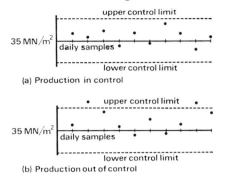

(a) Production in control

(b) Production out of control

quality control charts of a manufacturer (*see* Fig 4) can yield strong evidence as to the consistency of production.

Dimensions Attempts to design in multiples of brick sizes, to avoid cutting, do not always work out on site. Bricks below the intended size, but within the permitted tolerances, may require unsightly thick joints. Bricks above the intended size can be even more troublesome. The Standard suggests that 'where for special reasons closer tolerances are required . . . this can best be done by agreement between the user and the manufacturer and a useful basis of decision will be the routine control charts of brick dimensions'.

Design

Dampness Bricks differ in their tolerance of wet conditions but wet brickwork introduces risks of sulphate attack on mortar, efflorescence and frost damage. Brickwork should, therefore, be designed so that it does not remain excessively wet. This calls for correctly located and effective damp-proof courses, copings, flashings etc (*see* Digest 77).

For brickwork that is liable to be extremely wet for long periods, for example, parapets, free-standing and retaining walls, and chimney stacks, bricks of special quality are preferred. Parapets are seldom trouble-free and, if used, will often dictate the choice of brick for the whole of the external walling, as it will not usually be acceptable architecturally to use one facing up to roof level and a different one for the parapets.

Movement Thermal and moisture movements are discussed on page 1. The moisture expansion of brickwork is, on present evidence, about three-fifths that of the brick; the difference is probably due to restraint, as well as to mortar creep and shrinkage. BS 3921 does not provide a test for moisture expansion. Some manufacturers can provide useful data, but in the absence of this it is suggested that movement joints should provide for a *movement* of 10 mm in a 12 m length of walling. This will accommodate some movement due to sulphate attack but not all of the movement caused by severe sulphate attack. If the brickwork is effectively enclosed within a concrete or reinforced concrete frame, the combined effects of brickwork expansion and frame shrinkage and creep should be considered and appropriate compressible joints should be incorporated at tops, and in some cases at the ends, of the walls or infill panels. Short returns, under 600 mm long, of halfbrick thickness are specially vulnerable to cracking as a result of expansion (*see* Digest 75) and should be avoided.

Sulphate attack The risk of sulphate attack can be minimised by providing a generous overhang at eaves and verges, together with flashings and damp-proof courses at sills and elsewhere to prevent the ingress of water from above.

Parapets and free-standing walls should be avoided if possible; if not, these should be built to the recommendations of CP 121 Clause 3.5.3.6, taking the following precautions:

—design copings with generous overhang and adequate drip, with damp-proof course under.

—provide a continuous damp-proof course at roof level in parapets and at the base of free-standing walls, above the expected soil level. To develop bond so as to resist overturning, the latter could be of damp-proof course bricks set in 1:3 cement and sand mortar.

—provide movement joints not more than 12 m apart.

—use low-sulphate bricks.

—specify mortar mixes according to the previous recommendations, preferably using sulphate-resisting cement.

Earth-retaining walls are most vulnerable to attack and should be built in clay brickwork only if the bricks meet the requirements of BS 3921 for 'special' quality; sulphate-resisting mortar should be used. If, however, cavity construction is adopted, 'ordinary' quality bricks may be acceptable for the outer leaf (*see* Digest 89). Adequate copings and expansion joints should be provided. Drainage holes should be formed through the base of the wall; these must be lined if necessary to avoid conducting water into any cavity present.

Weathering

Natural weathering will usually enhance the appearance of good brickwork but some of the effects of poor detailing and design on appearance are discussed and illustrated in Digest 46.

Cleaning and maintenance

Routine maintenance and cleaning of brickwork is not normally required but if after prolonged exposure to a heavily polluted atmosphere cleaning becomes necessary, the methods described in Digest 113 can be used. The control of lichens, moulds and similar growths is discussed in Digest 139.

Some measures to be taken in the renovation of buildings damaged by floodwater are described in Digest 152.

Should repointing become necessary, the old mortar could be raked out to a depth of at least 20 mm and the mortar used for repointing should not be appreciably stronger than the original bedding mortar.

Ordering

When ordering, bricks should be identified by:

> variety – common, facing
> quality – internal, ordinary, special, engineering*, damp-proof course
> type – solid, perforated, hollow, cellular
> size – *see* Fig 1 and Table 2†

If 'internal' quality bricks are required for fair-faced work, this should be stated.

If appropriate, the strength class (*see* Table 5†) should be stated; this is essential for calculated load-bearing brickwork.

If required for facings, the colour and texture should be stated (usually by reference to agreed samples).

---

*Although defined in BS 3921 as a 'variety' this might more properly be considered as a 'quality'.
†Of Digest 164.

Now answer the following questions.

1　Clay bricks are classified in which British Standard?
2　Name the three varieties of bricks available.
3　Name the three qualities of clay brick available. Indicate their use.
4　Define 'early frost hazard'.
5　State how efflorescence occurs.
6　Name two situations where efflorescence commonly occurs.
7　How can sulphate attack be distinguished from moisture expansion without resorting to laboratory investigation?
8　Name the two qualities of clay brick which can be assumed to contribute to sulphate attack, since the amount of soluble sulphates is not limited according to BS 3921.
9　What mortar mixes are recommended to minimise the effect of sulphate attack?
10　State the suggested position of movement joints in a clay brick wall.

11   Draw up the following table and fill it in showing location of brickwork, the minimum quality and mortar mix recommended.

Cement: lime: sand and cement: sand with plasticiser.

Constructional element	No early frost hazard		Potential early frost hazard	
	Brick	Mortar mix	Brick	Mortar mix
Internal walls or inner leaf of cavity wall.				
Backing to solid wall. External.				
External Walls, outer leaf of cavity, (i) above d.p.c. (ii) below d.p.c.				
Parapets. Domestic chimney not rendered.				
Sills, copings and earth retaining walls.				

12   What are the minimum qualities of brick for various positions shown in the diagram?

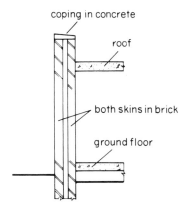

Calcium silicate bricks

Another type of brick is the calcium silicate brick. Read through carefully BRE Digest 157 *(Calcium silicate brickwork)*.

Building Research Establishment Digest 157

Calcium silicate (sandlime, flintlime) brickwork

Identification

Generic name

Calcium silicate bricks consist of a uniform mixture of sand, or uncrushed siliceous gravel, or crushed siliceous gravel or rock, or a combination of these, with a lesser proportion of lime, mechanically pressed and the materials chemically bonded by the action of steam under pressure. Suitable pigments may also be included.

Sandlime describes bricks in which only natural sand is used with the lime.

Flintlime bricks contain a substantial proportion of crushed flint.

Classes

BS 187: Part 2 specifies classes of calcium silicate bricks according to their compressive strength and drying shrinkage (*see* Table 1). The class numbers are a legacy of the values for compressive strength previously quoted in imperial units, viz compressive strength = class number $\times$ 1000 lbf/in^2 (approx).

Uses and limitations

Subject to selection of the appropriate class of brick and mortar, these bricks can give satisfactory service in a wide range of conditions of loading, dampness and freezing, internally and externally, above and below ground (*see* Fig 1 and Table 1). Most manufacturers offer sandlime and flintlime bricks in many colours in each of several classes.

All calcium silicate bricks, except those of Class 1 (which should not be used in wet conditions), are resistant to attack by most sulphate salts in soil and groundwater but where these are present the vulnerability of the mortar to attack must be considered. The bricks may be attacked by high concentrations of magnesium sulphate or of ammonium sulphate sometimes present in industrial wastes used for filling sites. If contaminated by chlorides, eg calcium chloride, common salt or seawater, all classes are liable to suffer if attacked by severe frost.

The drying shrinkage of calcium silicate bricks calls for care in design, in storage and during building – unless they are to be used in permanently damp conditions.

Fig 1   Minimum classes of calcium silicate bricks for various positions

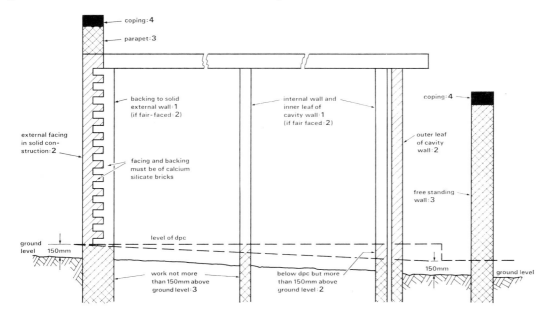

Table 1

CLASS	BS 187 : Pt 2 requirements		Suggested uses (it is permissible to use a stronger class of brick than is indicated)		Preferred mortar group for class of brick	
	Average compressive wet strength _MN/m²_	Maximum drying shrinkage _% of original wet length_			Frost risk during construction	
					none	_present_
1	7·0	Not specified	Internal walls	Not available in facing quality	v or iv	iv (air-entrained) or iii
			Inner leaf of cavity walls		iv	iii
			Backing to solid external walls			
2	14·0	2A 0·025 / 2B*0·035	External walls, outer leaf of cavity walls and facing to solid walls : work below dpc but only if more than 150 mm above ground level		iv	iii
3	20·5	3A 0·025 / 3B*0·035	Parapet walls, rendered	Brickwork liable to be continuously saturated with water (but not chloride solutions, _see_ text) or to be subjected to freezing when saturated	iv / iii	iii
			Parapet walls not rendered; external free-standing walls and work below 150 mm above ground level			
4	27·5	0·025	Sills and copings; earth-retaining walls back filled with free-draining material	Unprotected brickwork below ground—but if sulphates are present sulphate-resisting mortar is required	ii	ii
5	34·5	0·025	Calculated load-bearing brickwork wherever these high strengths are essential		to be specified by designer	
7	48·5	0·025				

*Not to be used where the high drying shrinkage would increase the risk of crack formation. Suitable therefore for piers, unrestrained or partially restrained infill panels of short length and short internal walls not restrained at both ends. Not to be used with mortars stronger than group iv—except that group iii may be used when there is risk of freezing.

Table 2 Mortar mixes (proportions by volume)

	Mortar group	Cement : lime : sand	Masonry-cement : sand	Cement : sand, with plasticiser
Increasing strength but decreasing ability to accommodate movements caused by settlement, shrinkage, etc	i	1 : 0–$\frac{1}{4}$: 3	—	—
	ii	1 : $\frac{1}{2}$: 4–4$\frac{1}{2}$	1 : 2$\frac{1}{2}$–3$\frac{1}{2}$	1 : 3–4
	iii	1 : 1 : 5–6	1 : 4–5	1 : 5–6
	iv	1 : 2 : 8–9	1 : 5$\frac{1}{2}$–6$\frac{1}{2}$	1 : 7–8
	v	1 : 3 : 10–12	1 : 6$\frac{1}{2}$–7	1 : 8

Direction of changes in properties

←——— equivalent strengths within each group ———→

increasing frost resistance ———→

improving bond and resistance to rain penetration ———→

Where a range of sand contents is given, the larger quantity should be used for sand that is well graded and the smaller for coarse or uniformly fine sand.

Because damp sands bulk, the volume of damp sand used may need to be increased. For cement : lime : sand mixes, the error due to bulking is reduced if the mortar is prepared from lime : sand coarse stuff and cement in appropriate proportions; in these mixes 'lime' refers to non-hydraulic or semi-hydraulic lime and the proportions given are for lime putty. If hydrated lime is batched dry, the volume may be increased by up to 50 per cent to get adequate workability.

Description

Constituents

Sand and flint aggregates are the major constituents and their suitability depends on the grading, the nature of the surface of the grains and any coatings on them, and the amounts of impurities present. The properties of the aggregates have an important influence on the quality of the bricks.

Either quicklime or hydrated lime is used but hydration must be complete before the bricks are pressed, to avoid expansion during subsequent steam treatment.

Suitable pigments can be incorporated in the mix and a high proportion of the bricks now made in this country are coloured in this way.

Manufacture

Mixing processes vary according to whether ground quicklime or hydrated lime is used. In the *silo or reactor process* ground quicklime and sand are mixed, usually in a screw or pan mixer, with an excess of water beyond that required to hydrate the lime; the mix is then stored in the silo for 3 to 24 hours to hydrate before being remixed and passed to the press. In the *drum hydration* process, ground quicklime and sand are mixed with a small excess of water over that required for hydration of the lime in a large revolving drum, in the presence of low-pressure steam which increases the temperature and so accelerates hydration. A pan mixer, or a rod mill, is then used to complete the mixing and to incorporate the remaining water necessary for pressing. In processes using dry hydrated lime, the sand and lime are mixed directly in the mixer or mill with the necessary water for pressing.

The bricks are then shaped under high mechanical pressure and put into hardening chambers (autoclaves) into which steam is slowly admitted until the required pressure is attained. The steam pressure is maintained for 4 to 15 hours according to the steam pressure used.

Associated materials

Mortar mixes for use with all classes of brick in various situations are set out in Table 2 (*see also* Digest 58).

Calcium silicate bricks do not impose any restrictions on the selection of materials to be used for damp-proof courses, flashings, ties, metal fastenings etc with which they might be in contact.

Shape and size

Accurate shape and size and square arrises are normal with this type of brick.

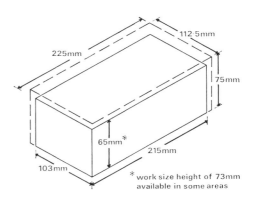

Fig 2 Work sizes (firm lines) and co-ordinating sizes (broken lines) of bricks to BS 187: Part 2

*work size height of 73mm available in some areas

The work sizes and co-ordinating sizes (ie inclusive of mortar joint) of bricks to BS 187: Part 2 are shown in Fig 2. The Standard specifies the following limits of manufacturing sizes:

	maximum *mm*	minimum *mm*
length	217	101
width	105	101
height	67	63

Bricks of co-ordinating sizes are available from some manufacturers. The sizes are likely to follow the recommendations of DD . . . *Clay bricks with modular dimensions* (in course of preparation) viz:

Designation	Work sizes length	width	height
mm	*mm*	*mm*	*mm*
300×100×100	290	90	90
200×100×100	190	90	90
300×100×75	290	90	65
200×100×75	190	90	65

Weight

Calcium silicate brickwork as laid weighs about 2 tonnes per cubic metre ($= 2$ kg/m² per mm thickness).

Appearance

The basic colours are white, off-white, cream or pale pink but facings are available in various pastel shades and darker hues. The white bricks in particular provide moderately good light reflectance when laid flush as facings to internal surfaces or to light wells and enclosed courts, without further applied decoration. All the colours deepen considerably when wet.

The surfaces of sandlime bricks are usually smoother than clay bricks but textured facings are available from some makers.

All bricks except those of Class 1 are required by the Standard to be free from visible cracks and noticeable balls of clay, loam and lime.

Characteristics

Strength

The strengths given in Table 1 are average compressive strengths when wet; the strength of dry calcium silicate bricks is 30–50 per cent higher than when saturated.

Fire resistance

Schedule 8 of the Building Regulations 1972 groups together clay, concrete and calcium silicate brickwork as having the same order of fire resistance.

Gases

The only gases normally present in the atmosphere which affect calcium silicate bricks are sulphur dioxide and carbon dioxide. Prolonged exposure to moist air containing sulphur dioxide decomposes the hydrated calcium silicate cementing agent and forms finally a skin of calcium sulphate (gypsum) and hydrated silica. Above ground, atmospheric carbonation, which occurs in depth slowly with age, slightly increases the strength but causes slight shrinkage.

Liquids

Absorption The total water absorption is usually 7–16 per cent (by weight), similar to some clay, other than engineering, bricks.

Moisture movements The drying shrinkage of bricks to BS 187: Part 2 should not exceed 0·025 per cent of the original wet length, except for bricks of classes 2B and 3B for which the limit is 0·035 per cent and class 1 for which no limit is specified (*see* Table 1). Subsequent expansion on wetting is likely to be about the same.

Thermal

Thermal conductivity The thermal conductivity (k value) of masonry of density 2000 kg/m^3 (from IHVE Guide) is:

at 1% moisture content* (= 'protected' situations):
0·92 W/m °C

at 5% moisture content* (= 'exposed' situations):
1·24 W/m °C

*moisture content by volume

Thermal expansion The coefficient of linear expansion is about 8 to 14×10^{-6} per °C, which is about one-and-a-half times that of fired clay bricks. This value is for the material in the unrestrained condition and for an area of walling it will be rather less.

Durability

The resistance of calcium silicate bricks to frost action is related mainly to their mechanical strength and this has been taken into account in making the recommendations set out in Table 1 and Fig 1.

They normally contain no soluble salts such as produce efflorescence on some clay bricks.

Exposure to salt spray causes some erosion of the surface, probably as a result of the repeated crystallisation of the salt with which they become impregnated, with alternate wetting and drying. They are not recommended for use where liable to exposure to sea-spray.

Working characteristics

The bricks can be cut readily to form closers etc but purpose-made bricks to BS 4729: 1971 can be obtained for many special purposes and, when available, should be used particularly in facing work.

The bricks do not take nails or screws directly, but they are easily drilled to receive plugs for these; they will quickly take the edge off masonry drills unless the detritus is cleared regularly from around the drill point.

Design

Movement joints

Below damp-proof course Brickwork below damp-proof course usually remains fairly wet; moisture movement, if any, is very slight and movement joints are not normally required.

External walls Vertical movement joints should be provided above damp-proof course at intervals of 7·5–9 m, subject to the shape of the panel and the disposition of any openings. The lines at which changes in height or thickness occur are lines of potential cracking; the higher the proportion of the length of a panel to its height, the greater is the risk of cracks developing in it.

Permanent movement joints, to remain in place throughout the life of the building, are formed by butting the bricks against a suitable separator, eg a 10 or 13 mm thick strip of joint filler, or bituminous felt, polythene sheeting or building paper, and pointing with a suitable mastic or sealant of matching or contrasting colour. They are always visible since, unless they have been constructed in toothed form, they break the bond but it is sometimes possible to locate them so that they are to some extent masked by an architectural feature. Where the use of non-setting mastic is undesirable, for example where it might be subject to deliberate damage, the joint may be pointed with mortar but this should be done as late as possible in the construction of the building. Alternatively, a self-curing sealant may be used but this may be more expensive; if so, its use is likely to be confined to the lowest one or two metres of walling.

An alternative method of forming a straight vertical joint is to leave a 10 mm gap severing the wall. A convenient way of doing this is to build in three thicknesses of 3 mm plywood or hardboard, the middle strip being wider and left projecting to allow for easy removal and release of the other two. The gap is then sealed with a preformed strip mastic or with pre-compressed foamed plastics.

Probably the least conspicuous method is to incorporate a continuous strip of clear polythene sheeting, say of 250 gauge, and of the same width as the thickness of the brick leaf or wall, following the mortar joints in toothed fashion.

Temporary movement joints may be left during construction, until the initial 'settling down and drying out' period is over. The joints are toothed to follow the bond or lap of the brickwork and formed with a mortar of low strength, eg a 1:3 to 1:4½ mix (by volume) of non-hydraulic lime and sand. After drying and settling down, the lime-sand mortar joints are deeply raked out and repointed with the same mortar as used in the rest of the work.

Permanent joints are preferred if the brickwork is to be pointed as the work proceeds.

Internal walls Internal walls built in mortar mix v (Table 2) do not require movement joints provided that they can be allowed to dry out before plastering. If, however, they must be plastered before the brickwork has dried out, or if they are built in a stronger mortar, movement joints should be provided.

Stability Care should be taken to ensure the stability of walls, especially those straight on plan, both during and after construction. Where necessary, sleeved dowels of constant cross-section, or other suitable fixings, may be used to span the movement joints, giving lateral support without restraining movement in the plane of the wall.

Junctions with other walling materials

Where calcium silicate bricks are to be used for one face of a solid wall, the bricks used for backing and for the other face should also be of calcium silicate or other material of comparable drying shrinkage and not clay bricks.

Where clay bricks are to be used in conjunction with calcium silicate facings, it is necessary to cater for differential movement between the two materials by constructing the wall in two leaves. These may be tied together with the normal complement of butterfly type wall ties but they must be separated, including all edges, with, for example, polythene sheeting or a cavity. If the cavity is to be narrower than 50 mm, special measures should be taken to avoid filling it with mortar. A separating membrane should be provided around all openings. The vertical damp-proof course of bituminous felt or polythene normally used in these positions in cavity external walls will generally perform this function but it is necessary to separate the two walling materials even where a vertical damp-proof course is not otherwise required. Cavity construction has a number of advantages for external walls, including better resistance to the weather; irrespective of the type of brick used, even a one-and-a-half brick thick solid wall, unless it is rendered externally or clad, will not be sufficiently resistant to rain penetration except in sheltered conditions.

All internal walls of calcium silicate brickwork, even those on upper floors, should be bedded on a damp-proof course or a strip of polythene sheeting. At all levels where clay bricks are bedded on calcium silicate brickwork, a similar measure should be taken so as to allow small longitudinal sliding movements to occur without undue restraint.

New brickwork should not generally be toothed and bonded to existing; a straight vertical movement joint, as previously described, should be interposed. This will allow the normal differential movement between the old work and the new to occur without adverse effect.

Junctions with concrete

Concrete-encased or reinforced concrete beams, lintels or slabs should not be bedded or cast directly on to calcium silicate brickwork, particularly facing brickwork, since the restraint afforded by the concrete may encourage the brickwork to crack, or differential movement between the concrete and the brickwork may cause a ragged horizontal crack at the junction. The two materials should be separated by, for example, two thicknesses of polythene sheeting; the special polythene sheeting made to give a low coefficient of friction is particularly suitable for this. For the slip plane to be fully effective, the top of the wall should be flushed up with mortar, trowelled smooth and allowed to harden, before the slip plane and concrete are placed. In external walls, a waterproof membrane at this level will also reduce or eliminate temporary darkening of the brickwork by wetting and any staining by lime or other substances that may be leached out of the concrete by the weather. A further precaution against these effects is to place a projecting drip of a stiff material (which may extend to the full width of the wall or leaf to act as the slip plane) at the junction; this will also help to prevent grout runs when casting *in situ* concrete. If considerations of lateral stability so dictate, the slip plane may be pierced at intervals and bridged vertically by butterfly type wire wall ties.

It is not advisable to place, or cast *in situ*, concrete encased or reinforced concrete columns directly against calcium silicate brickwork, or even to 'butter' the brickwork against the columns with mortar. The two materials should be separated by a slip plane, as described above.

When building up to the underside of beams or slabs, bituminous felt or some similar material should be interposed. If necessary for lateral stability, the sheeting may be pierced at intervals and bridged by wire ties, or some other suitable measures taken. For example, to assist with lateral restraint at the tops of walls, channels a little wider than the thickness of the wall can be formed in the concrete soffits and the brickwork built into them. Alternatively, dovetailed masonry fixings, consisting of proprietary slots and anchors, could be used. The metal slots should be cast into the concrete, so as to allow movement in the direction of the wall length, and the anchors placed at about 600 mm (maximum) centres. The anchors can be inserted in the slots as bricklaying proceeds, taken through the slip plane and bedded in the vertical joints of the brickwork.

Copings

If a brick-on-edge coping is to be used at the top of an external wall, a continuous damp-proof course of stiff material should be bedded immediately under the coping and should project to form a drip 25 mm or more beyond each face of the wall. A concrete coping (as BS 3798), preferably wide enough to oversail each face of the brickwork by about 50 mm and drip

throated, used in conjunction with a damp-proof course, will give better protection than a brick-on-edge coping.

Fig 3 Copings

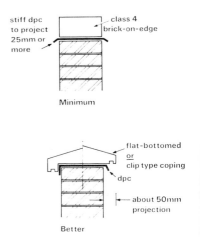

stiff dpc to project 25mm or more — class 4 brick-on-edge

Minimum

flat-bottomed or clip type coping

dpc

about 50mm projection

Better

Applied finishes

Tiling, plastering and rendering should not be carried over permanent movement joints, or across any junction with a different material where relative movement is to be expected. At these points each coat of plastering or rendering should either be severed by a straight cut, before the work hardens, or worked up to each side of a batten, temporarily or permanently secured at the edge of the joint. The joints can be covered by battens and strips (fastened to one side only of the joint) or by patent coverings, or filled with a sealant or mastic. Similar measures should be taken with wall tiling.

Sitework

Storage

To minimise drying shrinkage, the bricks should be reasonably dry when laid. Where manufacturers take steps to ensure delivery in such a condition, the benefit will be largely lost if the bricks are allowed to become wet on the site. They should, therefore, be stacked under cover or protected from rain by covering with polythene sheeting or other waterproof material.

Laying

Individual makes and classes of brick vary in their suction and absorption characteristics and the craftsman must adjust his technique accordingly. Bricks of high suction may need to be dampened before laying, particularly during hot weather, to prevent too much water being removed from the mortar; or, better, the consistency of the mortar should be adjusted to suit the suction of the units, if necessary by using water-retaining admixtures, rather than wetting the units to suit the mortar.

Protection after laying

Partly completed work, left overnight or longer, should be protected from rain.

Cleaning and maintenance

Grease, oil and general dirt can be removed with a scrubbing brush and warm water containing detergent.

Ordering

When ordering, the purchaser should specify:
   class
   size
   whether required for facing
   colour
   texture
   if certification to BS 187: Part 2 is required

Answer the following questions.

1   Describe the composition of calcium silicate bricks.
2   Define the brick type
   (a) sandlime   (b) flintlime
3   What is the British Standard for calcium silicate bricks?
4   Name the only class of calcium silicate brick which is *not* resistant to sulphate attack.
5   Draw up this table and fill it in showing location of walls to be built in calcium silicate bricks the minimum recommended class of brick and mortar (cement lime sand and cement with plasticiser).

Location	Class of calcium silicate brick	recommended mortar mix	
		frost hazard	no early frost hazard
Internal walls, outer leaf of hollow walls			
Backing to solid external walls			
External walls			
Parapet walls not rendered			
Sills and earth retaining walls			

6   When are movement joints necessary for walls in calcium silicate bricks? Indicate at what intervals.

Review

Before starting to build the wall, let's check on our learning so far.

What affects the price of bricks selected? Sketch a plan of the wall column junction. The frame is in structural steelwork, and the wall in a cavity construction.

13 SCAFFOLDING

Temporary working platforms are often required. In this section we will deal with tubular steel scaffold suitable for framed buildings with panel infill.

COMPONENT PARTS

This is an elevation of a scaffold.

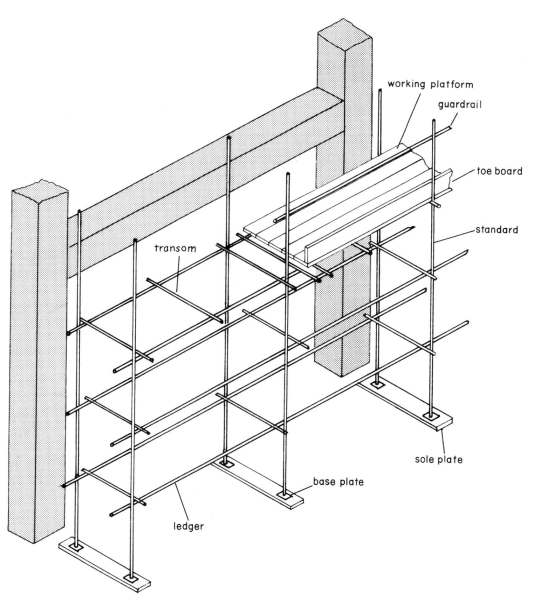

Can you label the following components of an independent scaffold: the **standard**, **ledger**, **transom** *and* **sole plate**?

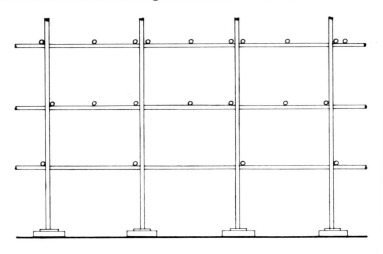

Sometimes an independent scaffold is called a double scaffold since it has two rows of standards.

THE WORKING PLATFORM

We will consider the working platform first. The Construction (Working Places) Regulations and the Health and Safety at Work Act lay down necessary safety precautions to be adopted in scaffolding. Let us look at the working platform, that is, the boarded out part of the scaffold.

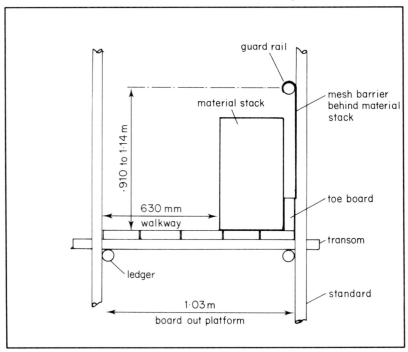

The Construction (Working Places) Regulations state the following widths of platform.

1 For walkway *only* 630 mm (that is, 3 boards wide)
2 For walkway and materials 830 mm (that is, 4 boards wide)
3 If barrows are used then add 200 mm to the 4 board platform. It is usual to use a 5 board platform giving a width of platform of 1030 mm.

If the working platform is 2 m above the ground it must have a **guard rail** and **toeboards**.

Each working platform board must be adequately supported along its length. The Construction (Working Places) Regulations stipulate the thickness of the board and the spacing of the transoms supporting the boards.

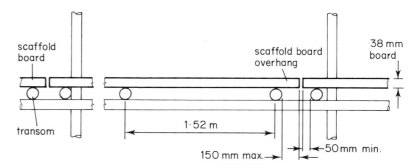

The transom supporting the working platform is positioned on to a tubular scaffold called a ledger. The ledgers are connected to the vertical members called standards with 90° coupling fittings.

Double coupler connects
tubes at 90°

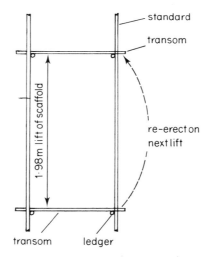

This 90° coupler would also be used to connect the transom to the ledgers. The distance between rows of ledgers is traditionally called a **scaffold lift**, and is about 1·9 metres purely to suit the scaffolder's reach.

STANDARDS

We can now turn our attention to the vertical members called standards. The Construction (Working Places) Regulations state that scaffold standards should be vertical and spaced close enough to support the ledgers properly. Steel base plates and sole plates of timber must be used to spread the scaffolding load over the ground to avoid possible local subsidence.

Independent tied scaffolds or double scaffolds can be classified into three types of scaffold according to their use.

Type of scaffold	Purpose or use
Light duty	For painting or cleaning facades of buildings. Only *one* working platform used at any one time.
General purpose	To provide up to four working platforms in use at any one time. For men and material stacks. Maximum distribution load per platform is 180 kg/m².
Heavy duty	Providing two heavy duty working platforms in use at any one time as well as two working platforms forms for access or light duty. Maximum distribution load per platform is 290 kg/m² and for access 180 kg/m².

The type of scaffold used will determine the spacing of the standards. We will detail only the general purpose scaffolding, as opposed to light duty or heavy duty.

On a general purpose scaffold up to 4 lifts may be boarded out at any one time. *What are the dimensions for*
    a) working platform width;
    b) height of guardrail and toe board?

The maximum spacings of the standards for general purpose scaffolding are indicated below.

guard rail

toe board

standard

boarded platform

ledger

maximum of 4 boarded platforms for this scaffolding

transom

base plate

sole plate

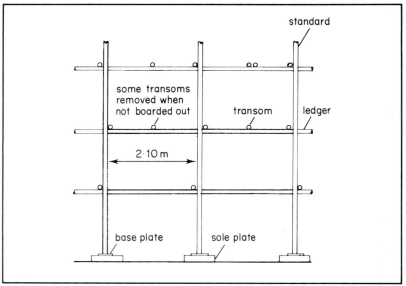

standard

some transoms removed when not boarded out

transom

ledger

2·10 m

base plate

sole plate

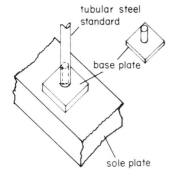

tubular steel
standard

base plate

sole plate

The load on the scaffold standard needs to be spread over a sufficient area of ground to avoid settlement. This is achieved by using a scaffolding fitting called a **base plate** and a timber **sole plate**.

Exercise

The builder wants a double scaffold to this building elevation. He proposes to hire the tubular scaffolding from a plant hire company at the rate of £0.10p per metre of tube hired.

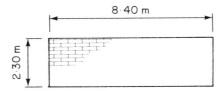

8·40 m

2·30 m

Elevation of building (assume scaffold
height is same as building height)

Calculate the amount of tubular scaffold required. Do not calculate for the transom tubes.

(See answer on p. 213)

STABILISING THE SCAFFOLDING

The erected tubular independent scaffold is not stable yet. It needs bracing and tying back to the structure.

Consider this elevation of an independent scaffold. The tendency is for the scaffolding to collapse. How can this be resisted?

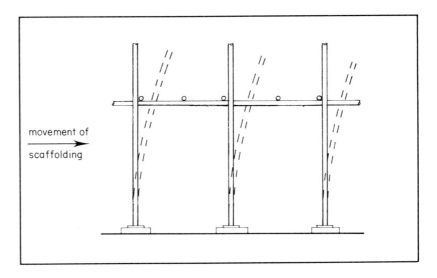

movement of
scaffolding

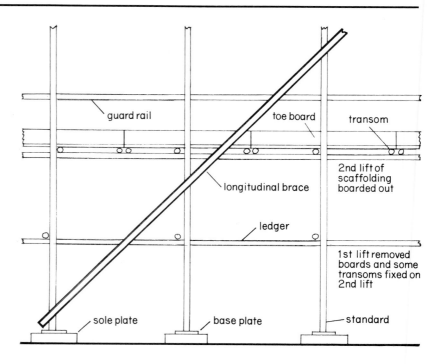

This is called **longitudinal bracing**. The angle of the raker should be at 45° and up to full height of the scaffold.

The standard fitting used to connect the raker to the standard is a swivel connector.

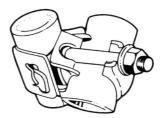

Swivel coupler connects tube
at any angle

The Construction (Working Places) Regulations stipulate that the scaffold should also be rigid across its width. To achieve this requirement **cross bracing** is used (sometimes called diagonal bracing).

We have introduced two new additional components necessary to an independent tied scaffold. Finally we must consider the fixing back of the scaffold to the structural frame.

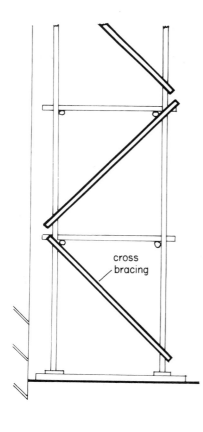

I CAN WELL REMEMBER THE HORRIBLE FEELING OF A SCAFFOLD MOVING AWAY FROM THE BUILDING. THANKFULLY THE INSURANCE PREMIUM WAS PAID UP AND THE SCAFFOLDING DID NOT MOVE FAR ENOUGH TO COLLAPSE.

How can we ensure that the scaffold does not move away from the building? *Draw a plan like the one below and indicate your possible solution.*

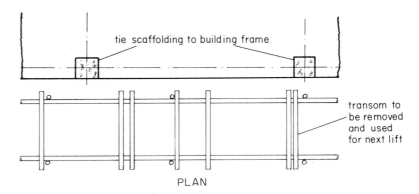

tie scaffolding to building frame

transom to be removed and used for next lift

PLAN

The recommendation is to secure the scaffolding to the building every 6 metres horizontally if possible, and to stagger these ties to the building.

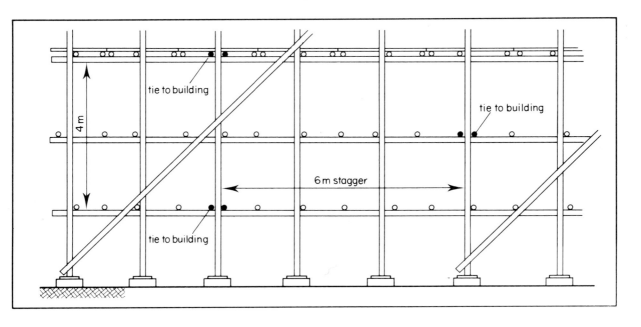

tie to building

4 m

tie to building

6 m stagger

tie to building

It would seem possible to use scaffold tube to yoke around the column. If that was your suggestion, well done. The detailing of this tie will be:

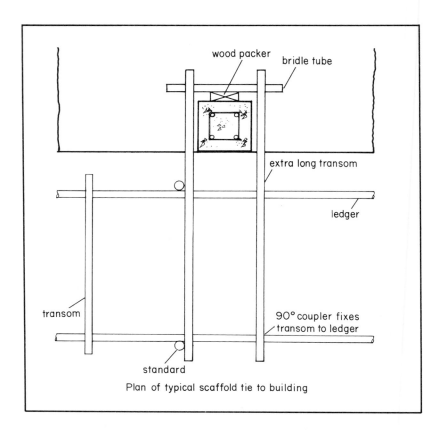

Plan of typical scaffold tie to building

Review

Now answer these questions.

1 What regulations cover the erection and dismantling of scaffolding?

2 Sketch a diagram(s) to show the location of the following scaffold components:
   a) standard; b) ledger; c) transom; d) guardrail;
   e) toeboard; f) longitudinal and cross bracing.

3 State the dimensions for:
   a) spacing of standards for a medium duty scaffold.
   b) spacing of a transom to support a 38 mm thick scaffold board.
   c) the maximum and minimum overhang of a scaffold board, beyond a transom.
   d) the distance between ties to a building for an independent tied medium duty scaffold.

14 PRECAST CONCRETE FLOORS

There are several advantages in using precast concrete floors rather than **in situ** concrete floors. The most obvious advantage is speed of erection and for this reason they are often used in conjunction with steel framed buildings. They can also be loaded immediately and there is no need for formwork during erection.

There are four main types of concrete floor: solid units, hollow units, trough or tee units, and beam and block floors.

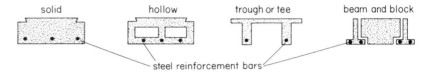

solid   hollow   trough or tee   beam and block

steel reinforcement bars

SOLID UNITS

Solid units either have traditional mild steel bar reinforcement or they are prestressed. *Find out what prestressed concrete is. What are its advantages over reinforced concrete?* The dead load of a solid unit precast concrete floor is approximately the same as an in situ concrete floor, unless the units are made using **auto claved aerated concrete**. (You will need to find out what that is before answering question 1 on page 126.)

The solid units are placed on the steel or concrete frame with the aid of continuity rods.

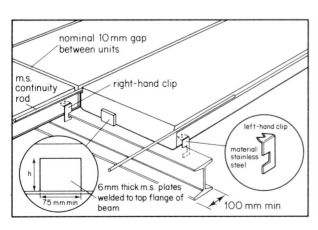

nominal 10 mm gap between units

m.s. continuity rod

right-hand clip

left-hand clip

material: stainless steel

6 mm thick m.s. plates welded to top flange of beam

h

75 mm min

100 mm min

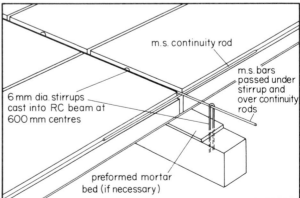

m.s. continuity rod

m.s. bars passed under stirrup and over continuity rods

6 mm dia. stirrups cast into RC beam at 600 mm centres

preformed mortar bed (if necessary)

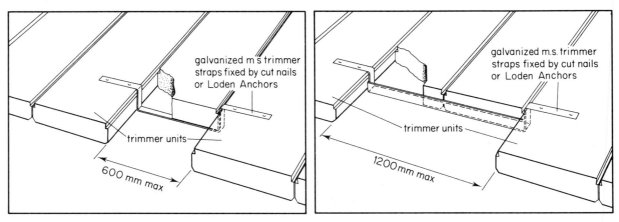

Openings are made using galvanised mild steel trimmers to support the ends of the units.

HOLLOW UNITS

Hollow units are manufactured as pre-stressed concrete units. **High yield steel** rather than mild steel is used. The steel bar is fixed at one end of the unit using an anchor plate and nut. The steel is then stretched, either whilst the concrete is still green or after it has hardened, then another anchor plate and nut are positioned and the stress induced in the steel is transmitted back to the concrete. *What stress is the concrete under now?*

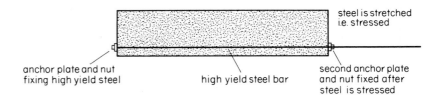

Hollow units are placed in two different ways.

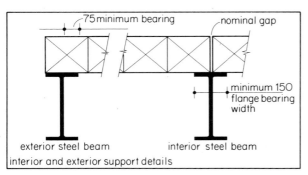

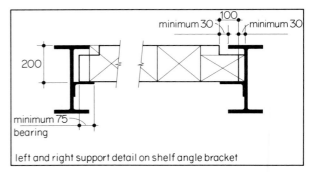

What is the advantage of using the shelf angle bracket method?

Freshly made concrete is used to flush up the joints and, where appropriate, to satisfy fire regulations.

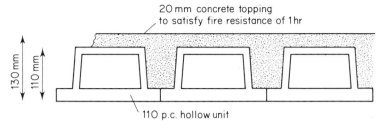

TROUGH OR TEE SECTIONS

Precast trough and tee sections allow a reduction of the floor dead weight whilst retaining the same load carrying capacity as the solid dense concrete floor. They can be used with concrete or steel framed buildings and typical fixing details are shown here.

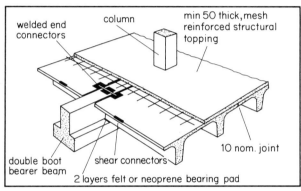

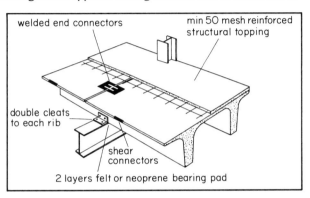

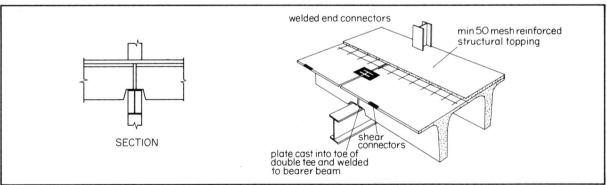

BEAM AND BLOCK

This type of floor is based on a well-tried principle, that of the timber floor, where you are likely to be completely safe with just 25mm of board under your feet. The beam and block floor utilizes this idea.

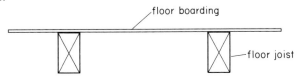

The beams are spaced apart, lightweight infill blocks are used to make the floor solid and a structural concrete topping is laid to finish the floor.

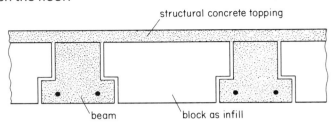

This is a composite flooring system, using both precast units and in situ concrete, and sometimes temporary supports under the floor are needed until the concrete matures.

The beam can have a top seating when used with either a steel or reinforced concrete frame, and is connected in the same way as the hollow floor unit. *Try sketching a section through the beam and floor to show a top seating on to a steel intermediate beam. The minimum seating is 75mm.* To reduce the overall height of the building the connection could also be to a shelf angle to a steel beam.

Remember: All precast concrete floors rely on cranes to lift and place the units. This must be considered when the builder plans his site layout.

Activity

1 Obtain the Building Research Establishment Digest No. 178 and answer these questions.
   a What is aerated concrete made from?
   b How is the cellular structure normally produced in aerated concrete?
   c What is autoclaving?
   d What does the autoclaving do for the aerated concrete?
   e What are the two main precast forms of autoclaved concrete?
   f In using reinforced autoclaved concrete units what problem caused most concern to the manufacturers?
   g List the four special advantages of using precast auto claved aerated concrete floor and roof units.
   h Why must care be taken in handling reinforced precast concrete aerated concrete floor units?
2 Sketch a detail showing trimming around a stairwell using a trough section precast floor, and an in situ concrete frame.
3 Explain the advantages of pre-stressed concrete floor units compared to mild steel bar reinforcements.
4 Obtain some trade literature giving span/load tables for one of the four types of precast concrete floors. Use the tables to calculate the number of floor units required for the first floor of the four storey office building shown here. The floor loading is 2·5 kN per m². Do not allow for the stair opening.

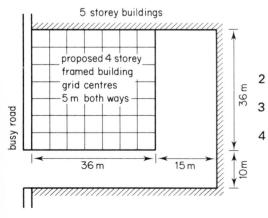

Stair sizes are covered by Building Regulations Part H *Stairways, ramps, balustrades and vehicle barriers*. Read through Part H3 and answer the questions which follow. Remember offices are Purpose Group IV.

Table to Regulation H3

Specific requirements for stairways

Head	Building or compartment of purpose group I or III–	Building or compartment of purpose group I or III–	Building or compartment of purpose group II or VII–	Building or compartment of purpose group II, III, IV, V, VI, VII or VIII–
	any stairway within a dwelling or serving exclusively one dwelling	any stairway for common use in connection with two or more dwellings	any stairway– (a) within or serving a building or compartment of purpose group II other than a stairway for use solely by staff; or (b) serving a part of a building or compartment of purpose group VII more than 100 m² in area and used for assembly purposes	any stairway other than a ramp to which either column (2), (3) or (4) relates
(1)	(2)	(3)	(4)	(5)
A. Width of stairway (subject to the provisions of Section II of Part E)	Not less than– (a) 600 mm in the case of a stairway providing access only to– (i) one room, not being a living room or kitchen; or (ii) a bathroom and a watercloset; or (b) 800 mm in any other case	Not less than 900 mm	Not less than 1 m	Not less than– (a) 800 mm in the case of a stairway within or serving a part of a building or compartment which is not capable of being used or occupied by more than 50 persons; or (b) 1 m in any other case
B. Additional requirement for stairways over 1.8 m in width	—	—	Each flight to be so subdivided into sections that each section is– . (a) not less than 1 m nor more than 1.8 m in width; and (b) separated from any other such section by a handrail complying with the requirements set out against head K	Each flight to be so subdivided into sections that each section is– (a) not less than 1 m nor more than 1.8 m in width; and (b) separated from any other such section by a handrail complying with the requirements set out against head K
C. Pitch of flight	Not exceeding 42°	Not exceeding 38°	—	—
D. Number of rises per flight. This requirement shall not apply to any step giving access to a dais, stage, shop window or a small room only or situated at an external doorway	Except at the bottom of a stairway, not fewer than 2 nor more than 16	Not fewer than 2 nor more than 16	Not fewer than 3 nor more than 16	Not fewer than 3 nor more than 16
E. Height of rise	Not less than 75 mm nor more than 220 mm	Not less than 75 mm nor more than 190 mm	Not less than 75 mm nor more than 180 mm	Not less than 75 mm nor more than 190 mm
F. Going of step (subject to the provisions of head J)	Not less than 220 mm	Not less than 240 mm	Not less than 280 mm	Not less than 250 mm

Table to Regulation H3 – continued

Specific requirements for stairways

Head	Building or compartment of purpose group I or III–	Building or compartment of purpose group I or III–	Building or compartment of purpose group II or VII–	Building or compartment of purpose group II, III, IV, V, VI, VII or VIII–
	any stairway within a dwelling or serving exclusively one dwelling	any stairway for common use in connection with two or more dwellings	any stairway– (a) within or serving a building or compartment of purpose group II other than a stairway for use solely by staff; or (b) serving a part of a building or compartment of purpose group VII more than 100 m² in area and used for assembly purposes	any stairway other than a ramp to which either column (2), (3) or (4) relates
(1)	(2)	(3)	(4)	(5)
G. Aggregate of the going and twice the rise of a step (subject to the provisions of head J). This requirement shall not apply to a flight which has only one rise	Not less than 550 mm nor more than 700 mm	Not less than 550 mm nor more than 700 mm	Not less than 550 mm nor more than 700 mm	Not less than 550 mm nor more than 700 mm
H. Going of landings (subject to the provisions of Section II of Part E)	Not less than the width of the stairway	Not less than the width of the stairway	Not less than the width of the stairway or (if the stairway is subdivided) the width of the wider or widest section	Not less than the width of the stairway or (if the stairway is subdivided) the width of the wider or widest section
J. Tapered treads	(a) The going of any part of a tread within the width of the stairway to be not less than 75 mm *(b) The going to be not less than 220 mm *(c) The aggregate of the going and twice the rise to be not less than 550 mm nor more than 700 mm *(d) The pitch to be not more than 42°	(a) The angle (measured on plan) formed by the nosing of the tread and the nosing of the tread or landing immediately above it to be not more than 15° *(b) The going to be not less than 240 mm *(c) The aggregate of the going and twice the rise to be not less than 550 mm nor more than 700 mm *(d) The pitch to be not more than 38°	(a) The angle (measured on plan) formed by the nosing of the tread and the nosing of the tread or landing immediately above it to be not more than 15° *(b) The going to be not less than 280 mm *(c) The aggregate of the going and twice the rise to be not less than 550 mm nor more than 700 mm	(a) The going of any part of a tread within the width of the stairway to be not less than 75 mm (b) The angle (measured on plan) formed by the nosing of the tread and the nosing of the tread or landing immediately above it to be, in the case of a stairway 1 m or more in width, not more than 15° *(c) The going to be not less than 250 mm *(d) The aggregate of the going and twice the rise to be not less than 550 mm nor more than 700 mm
	*For the purposes of (b), (c) and (d) above, the going, rise and pitch shall be measured at the central points of the length (or, where applicable, the deemed length) of a tread if the stairway is less than 1 m in width, or at points 270 mm from each end of the length (or where applicable the deemed length) of a tread if the stairway is 1 m or more in width	*For the purposes of (b), (c) and (d) above, the going, rise and pitch shall be measured at points 270 mm from each end of the length (or where applicable the deemed length) of a tread	*For the purposes of (b) and (c) above, the going and rise shall be measured at points 270 mm from each end of the length (or where applicable the deemed length) of a tread	*For the purposes of (c) and (d) above, the going and rise shall be measured at the central points of the length (or, where applicable, the deemed length) of a tread if the stairway is less than 1 m in width, or at points 270 mm from each end of the length (or, where applicable, the deemed length) of a tread if the stairway is 1 m or more in width

K. Handrails
These requirements shall not apply to any side of a flight formed by fixed seating
	Irrespective of the purpose group of the building or compartment–
	(a) any flight with a total rise of more than 600 mm shall be provided with a handrail–
	  (i) on each side of the flight if the width of the flight is 1 m or more;
	  (ii) on the side where the tapered treads have the greater going if the flight is less than 1 m in width and contains tapered treads; and
	  (iii) on at least one side in any other case; and
	(b) any such handrail shall–
	  (i) be so designed as to afford adequate means of support to persons using the flight;
	  (ii) be continuous for the length of the flight (except that any handrail need not extend beside the two steps at the foot of a stairway);
	  (iii) be securely fixed at a height of not less than 840 mm nor more than 1 m (measured vertically above the pitch line); and
	  (iv) be terminated by a scroll or other suitable means

1   How wide should a stairway serving 60 people be?
2   What is the maximum number of steps per flight?
3   What is the minimum and maximum step rise?
4   The aggregate of the going and twice the rise is to be not less than what; and not more than what?
5   Handrails must be incorporated where the total rise exceeds how many mm?
6   How many handrails must be included on a stair width of 1·2 metres?
7   The handrail must be securely fixed at a height above the pitch line. What is the minimum and maximum height?

CONSTRUCTION OF STAIR IN REINFORCED CONCRETE

If an opening is made in a reinforced concrete floor the tendency at the edges is for the slab to deflect.

length of stairwell

edge tends to deflect

deflection of edge

To stiffen the edges of the opening, edge beams are incorporated.

edge beam

beam width of

depth of beam = 1 1/2 width

What would be the typical steel reinforcement to the edge beam?

For learning purposes only, the stairs can be considered in two stages: first the slab and secondly the steps.

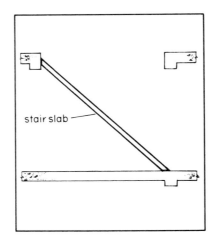 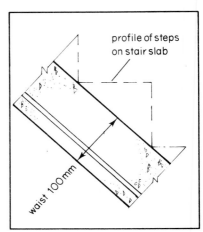

The slab spans between floor and landing supports, and will carry the dead and imposed loads on the stair to the floor. The thickness of the slab depends on load, span and width of stairs. The Code of Practice CP 110 states the minimum thickness of the waist as 100 mm.

The completed detail of the reinforced concrete stair slab is as follows:

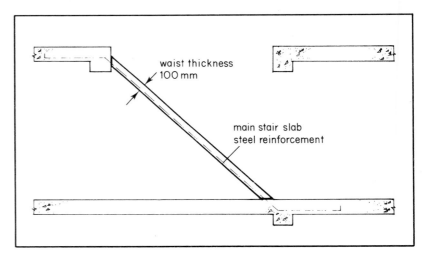

Without the steel reinforcement the slab will deflect when loaded.

Constructional sequence dictates that the stairs are cast after the concrete structure. The reinforcement of the stair slab needs to be spliced to the concrete floor steel reinforcement. Starter bars are cast into the floor slab. Later when the stairs are cast the stair slab reinforcement is spliced to these starter bars.

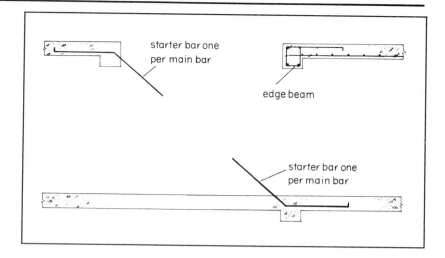

There is, of course, a danger in leaving starter bars protruding. *To avoid accidents the area should have barriers around.*

The splice length depends on whether the bars joined together are in compression or tension. For splicing steel in tension it is recommended that the splice length be 30 times bar diameter. The number of main reinforcement bars depends on the width of the stairs. For light office duties the stair main reinforcements could be at 150 centre to centre. For example if the stair width was 900 mm the number of bars would be:

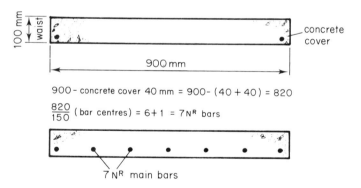

900 – concrete cover 40 mm = 900 – (40 + 40) = 820

$\frac{820}{150}$ (bar centres) = 6 + 1 = 7 N^R bars

7 N^R main bars

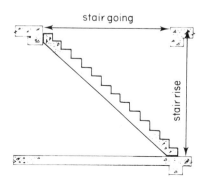

We are now in a position to consider the step details. There are several important dimensions to be incorporated from the Building Regulations.

Answer the following questions: What is the maximum rise? What is the minimum going? What is the maximum number of steps in one flight? Where should the main reinforcement to the stair slab and floor go?

Main reinforcement bars by themselves are inadequate – distribution steel needs to be incorporated. As a guide allow *one distribution bar per step.*

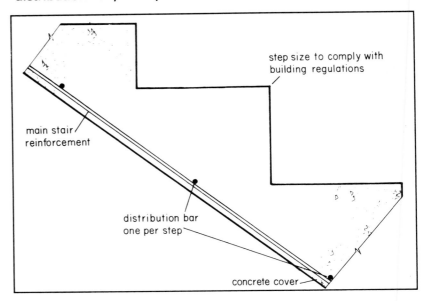

We have now completed the principles and components to a reinforced concrete *in situ* staircase. The working drawings used by the builder will include steel layout drawing showing plan and section in addition to the steel bending schedule.

Review

Calculate the number of main bars and distribution bars required for this stair going and rise.

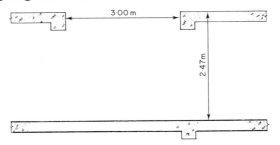

Attempt the work in this order:
1 Calculate number of steps within the Building Regulation maximum step rise.
2 Calculate the stair going (number of steps × minimum going).
3 Check pitch angle of stair to be within Building Regulations. Adjust going if necessary to suit Regulations.
4 The stairs are 900 mm wide, with concrete cover at edge of 40 mm. Calculate number of main bars.
5 Draw the diagram including the stair profile. Indicate position of steel reinforcement.

16 PROPRIETARY PARTITIONS

What was wanted? A partition. As you know, the difference between a partition and other types of walls is that it does not have to carry any structural loads. Partitions can be classified by materials used, or by their degree of moveability.

Remember brick and timber stud partitions? Which of these was easier to move? Obviously it is difficult to classify them in this way. That is why the industry in its wisdom classifies partitions by their degree of **demountability**. At least it is a longer word and therefore more impressive.

What is a demountable partition? Some of the traditional partitioning materials would take a sledge hammer to move. The folding or moveable screens compared with masonry (you know that as brick or blockwork) are easily moved. But what about that other type of partition, the demountable partition?

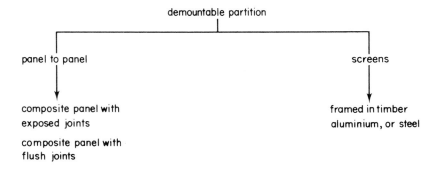

We will concern ourselves with just the **panel to panel** demountable partition.

Paramount is a good example of proprietary partitions. It is a panel/panel demountable partition manufactured by the British Gypsum Co Ltd. It is a composite panel of two sheets of plasterboard. Between these two sheets is a cellular core of stiff paper.

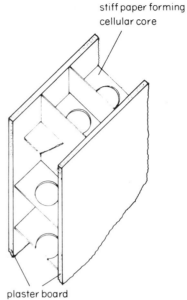

The Paramount panel is manufactured to BS 4022. *Prefabricated gypsum wall board panels.* What sizes? A range of sizes are available in thickness and panel size. The edges of the Paramount are finished in one of three ways.

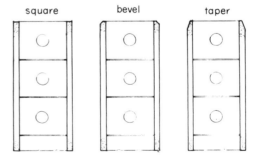

We will discuss only the taper edged board.

Paramount is light, but is it suitable as a partitioning material? It depends on what you want the partition to do or what its performance requirements are. You do not need load bearing ability since the structural frame will carry the floor loads . . . but you do need rigidity. To achieve rigidity Paramount partitioning has a timber frame.

There are other performance requirements that will have to be considered: sound insulation, possible accepting of services and fittings for instance.

Of great concern too will be fire insulation and flame spread.

ERECTING THE PARTITION

How is this demountable partition erected? For rigidity softwood framing is introduced into the partitioning. Using 50 mm thick Paramount the ceiling and wall battening is usually 30 mm × 19 mm. The intermediate battens between the panels are normally 30 mm × 37 mm, whereas the sole plate is 50 mm × 19 mm. At each panel to panel a 30 mm × 19 mm length of timber is nailed to the sole plate. This additional batten is 300 mm long.

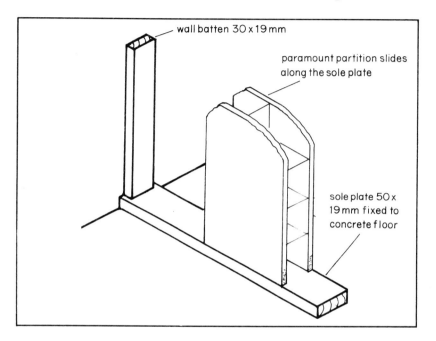

wall batten 30 x 19 mm

paramount partition slides along the sole plate

sole plate 50 x 19 mm fixed to concrete floor

The ceiling batten sits inside the Paramount.

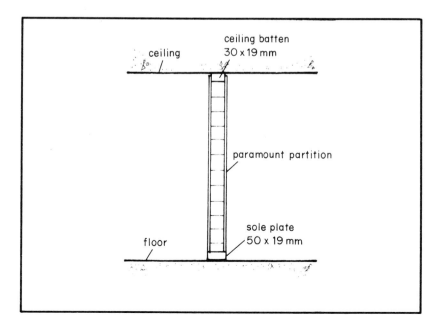

ceiling

ceiling batten 30 x 19 mm

paramount partition

floor

sole plate 50 x 19 mm

At the wall junction the Paramount batten is pushed over the wall batten.

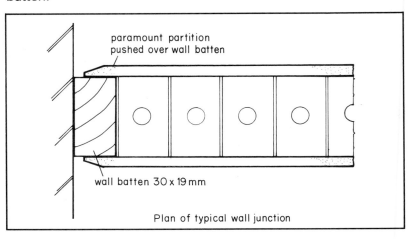

Plan of typical wall junction

What next? The Paramount is nailed to the battening using galvanised flat headed nails, 2 mm × 30 mm long. Next, the panel to panel joints can be flushed up.

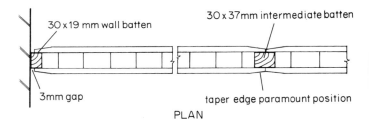

PLAN

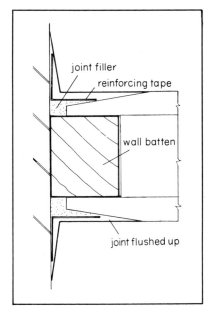

The panel is positioned over the sole plate and fixed to the wall batten leaving a gap of 3 mm. A special joint compound called **joint filler** is mixed, applied to the gap. Whilst the joint filler is still soft, paper tape is pressed into the joint filler.

This type of joint filling is the same for the ceiling joint.

The intermediate joint

Let us now consider the intermediate joint, that is, the panel to panel joint. The steps are exactly the same as for the wall/panel junction. Joint filler, press in the paper tape, then flush the joint. *There is, though, one difference; what is it?*

The joint is finally flushed up to make a smooth joint.

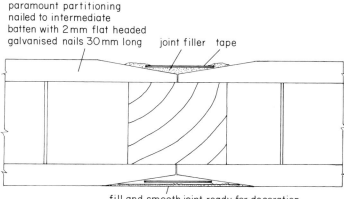

paramount partitioning nailed to intermediate batten with 2 mm flat headed galvanised nails 30 mm long   joint filler   tape

fill and smooth joint ready for decoration

The difference between an intermediate panel to panel joint and the wall to panel junction is that the boards are pushed right together at the intermediate, whereas the wall joint a gap of 3 mm is left, and the 300 mm long ×30 mm×19 mm batten nailed to the sole plate.

The demountable partition is now ready for decoration without the need of a skimming coat of plaster, traditionally used with the other type of remountable partitioning (the timber stud partition). That means the Paramount panel must be finished ready for decoration. Can you remember the two grades of plasterboard which are ready for decoration? On ceilings finished in the traditional way baseboard was used needing a skimming of wet plaster to make it ready for decoration. The alternative is wall board. Here the plasterboard is finished ready for decoration and only the joints needing finishing.

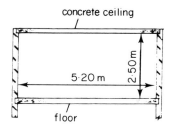

concrete ceiling

5·20 m   2·50 m

floor

Exercise

Being a fully fledged partition erector why not try your skill on this exercise?

The Paramount panel is 2·50 m long × 1·00 m wide. Calculate the length of battening required for this partition.

(See answer on p. 213)

The Paramount panel is 50 mm thick by 900 mm wide × 2·50 m long. Now, consider carefully and list the sequence of erecting the panels.

Thankfully most rooms need doors. If we insert a doorway in that partition will it make erecting easier?

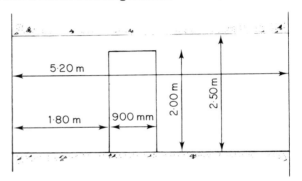

To avoid further embarrassment, work to the door jamb then start again at the other wall putting panels in up to the other door jamb.

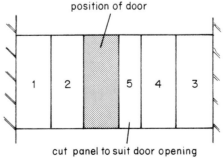

cut panel to suit door opening

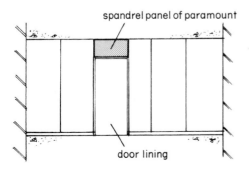

spandrel panel of paramount

door lining

What type of door? A door lining could be used with a panel of Paramount over the door head. This is called a **spandrel panel**.

Is there an alternative method? A storey height door lining could be used, it would solve the problem of fixing the spandrel panel of Paramount.

By the way, *how will the spandrel panel be fixed?* The existing battening to the Paramount must include a batten at the door jamb.

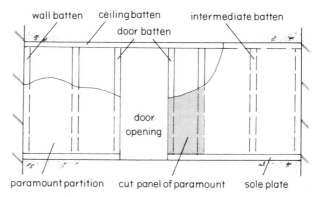

Here is a close view of the way in which the spandrel panel is fixed.

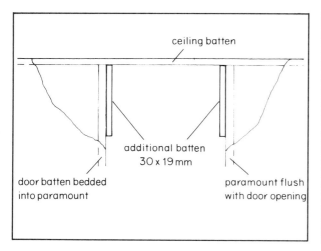

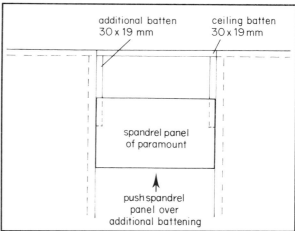

Once the battening is fixed the spandrel panel can be pushed up and over the battening. The completed partition looks like this.

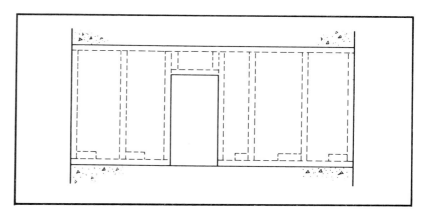

17 FIRE DOORS

Read through carefully Part E *Safety in Fire* section of the Building Regulations and answer the questions which follow.

PART E

Safety in fire

E1 Interpretation of Section 1

(1) In this Section and in the schedules thereto–

BASEMENT STOREY means a storey which is below the ground storey; or, if there is no ground storey, means a storey the floor of which is situated at such a level or levels that some point on its perimeter is more than 1.2 m below the level of the finished surface of the ground adjoining the building in the vicinity of that point;

CAVITY and CAVITY BARRIER have the meanings assigned by regulation E14(1);

COMPARTMENT means any part of a building which is separated from all other parts by one or more compartment walls or compartment floors or by both such walls and floors; and, if any part of the top storey of a building is within a compartment, that compartment shall also include any roof space above such part of the top storey;

COMPARTMENT WALL and COMPARTMENT FLOOR mean respectively a wall and a floor which complies with regulation E9 and which is provided as such for the purposes of regulation E4 or to divide a building into compartments for any purpose in connection with regulation E5, E6 or E7;

DOOR includes any shutter, cover or other form of protection to an opening in any wall or floor of a building or in the structure surrounding a protected shaft, whether the door is constructed of one or more leaves.

E9 Compartment walls and compartment floors

(1) Any compartment wall or compartment floor shall be imperforate with the exception of any one or more of the following–

(a) (i) in the case of a compartment wall separating a flat or maisonette from any space in common use giving access to that flat or maisonette, an opening fitted with a door which complies with the requirements of regulation E11 and has fire resistance of not less than half an hour; or

(ii) in any other case, an opening fitted with a door which complies with the requirements of regulation E11 and has fire resistance of not less than the minimum period required by regulation E5 for the wall or floor; or

(b) an opening for a protected shaft; or

(c) an opening for a ventilation duct (other than a duct in, or consisting of, a protected shaft) if any space surrounding the duct is fire-stopped and the duct is fitted with an automatic fire shutter where it passes through the wall or floor; or

(d) an opening for a pipe which complies with the requirements of regulation E12; or

(e) an opening for a chimney, appliance ventilation duct or duct encasing one or more flue pipes, in each case complying with the relevant requirements of paragraph (5) and of Part L; or

(f) an opening for a refuse chute which complies with the requirements of Part J.

E11 Fire-resisting doors

(1) This regulation shall apply to any door which is required by the provisions of this Section to have fire resistance.

(2) In this regulation–

AUTOMATIC SELF-CLOSING DEVICE does not include rising butt hinges except in relation to a door to which paragraph (5) applies; and

ELECTRO-MAGNETIC OR ELECTRO-MECH-ANICAL DEVICE SUSCEPTIBLE TO SMOKE refers only to any such device which will allow the door held open by it to close automatically upon the occurrence of each or any one of the following–

(a) detection of smoke by automatic apparatus suitable in nature, quality and location;

(b) manual operation of a switch fitted in a suitable position;

(c) failure of electricity supply to the device, apparatus or switch;

(d) if a fire alarm system is installed in the building, operation of that system.

(3) (a) Any door to which this regulation applies shall (subject to paragraph (7)) be fitted with an automatic self-closing device.

(b) No means of holding any such door open shall be provided other than a fusible link or, if the door is so constructed and installed that it can readily be opened manually, an electro-magnetic or electro-mechanical device susceptible to smoke.

(c) No part of a hinge on which any such door is hung shall be made either of combustible material or of non-combustible material having a melting point less than 800°C.

(4) Any door fitted in an opening which is provided as a means of escape in the event of fire or might be so used shall be so constructed and installed that it can readily be opened manually and shall not be held open by any means other than an electro-magnetic or electro-mechanical device susceptible to smoke:

Provided that there may also be installed so as to close the same opening a door which cannot readily be opened manually if–

(a) such door is fitted with an automatic self-closing device and is held open by a fusible link;

(b) the manually openable door has fire resistance of not less than half an hour; and

(c) the required fire resistance is achieved by the two doors together.

Schedule 8 (Regulation E1(5), proviso(a))   **Deemed-to-satisfy provisions**

Notional periods of fire resistance

In the following Table–

(a) CLASS 1 AGGREGATE means foamed slag, pumice, blast-furnace slag, pelleted fly ash, crushed brick and burnt clay products (including expanded clay), well-burnt clinker and crushed limestone; and
CLASS 2 AGGREGATE means flint gravel, granite and all crushed natural stones other than limestone;

(b) any reference to plaster means–
(i)   in the case of an external wall 1 m or more from the relevant boundary, plaster applied on the internal face only; or
(ii) in the case of any other wall, plaster applied on both faces; or
(iii) if to plaster of a given thickness on the external face of a wall, except in the case of a reference to vermiculite-gypsum or perlite-gypsum plaster, rendering on the external face of the same thickness; or
(iv) if to vermiculite-gypsum plaster, vermiculite-gypsum plaster of a mix within the range of $1\frac{1}{2}$ to 2:1 by volume; and
(c) in the case of a cavity wall, the load is assumed to be on the inner leaf only except for fire resistance period of four hours.

Part I: Walls   A. Masonry construction

Construction and materials	Minimum thickness excluding plaster (in mm) for period of fire resistance of–									
	Loadbearing					Non-loadbearing				
	4 hours	2 hours	$1\frac{1}{2}$ hours	1 hour	$\frac{1}{2}$ hour	4 hours	2 hours	$1\frac{1}{2}$ hours	1 hour	$\frac{1}{2}$ hour
1. Reinforced concrete, minimum concrete cover to main reinforcement of 25 mm:										
(a) unplastered	180	100	100	75	75					
(b) 12.5 mm cement-sand plaster	180	100	100	75	75					
(c) 12.5 mm gypsum-sand plaster	180	100	100	75	75					
(d) 12.5 mm vermiculite-gypsum plaster	125	75	75	63	63					

Construction and materials	Loadbearing					Non-loadbearing				
	4 hours	2 hours	1½ hours	1 hour	½ hour	4 hours	2 hours	1½ hours	1 hour	½ hour
2. No-fines concrete of Class 2 aggregate:										
(a) 12.5 mm cement-sand plaster						150				
(b) 12.5 mm gypsum-sand plaster						150				
(c) 12.5 mm vermiculite-gypsum plaster						150				
3. Bricks of clay, concrete or sand-lime:										
(a) unplastered	200	100	100	100	100	170	100	100	75	75
(b) 12.5 mm cement-sand plaster	200	100	100	100	100	170	100	100	75	75
(c) 12.5 mm gypsum-sand plaster	200	100	100	100	100	170	100	100	75	75
(d) 12.5 mm perlite-gypsum plaster (to clay bricks only)	100	100	100	100	100	100	100	100	75	75
(e) 12.5 mm vermiculite-gypsum plaster	100	100	100	100	100	100	100	100	75	75
4. Concrete blocks of Class 1 aggregate:										
(a) unplastered	150	100	100	100	100	150	75	75	75	50
(b) 12.5 mm cement-sand plaster	150	100	100	100	100	100	75	75	75	50
(c) 12.5 mm gypsum-sand plaster	150	100	100	100	100	100	75	75	75	50
(d) 12.5 mm vermiculite-gypsum plaster	100	100	100	100	100	75	75	62	50	50
5. Concrete blocks of Class 2 aggregate:										
(a) unplastered		100	100	100	100	150	100	100	75	50
(b) 12.5 mm cement-sand plaster		100	100	100	100	150	100	100	75	50
(c) 12.5 mm gypsum-sand plaster		100	100	100	100	150	100	100	75	50
(d) 12.5 mm vermiculite-gypsum plaster	100	100	100	100	100	100	75	75	75	50
6. Autoclaved aerated concrete blocks, density 475–1200 kg/m³	180	100	100	100	100	100	62	62	50	50
7. Hollow concrete blocks, one cell in wall thickness, of Class 1 aggregate:										
(a) unplastered		100	100	100	100	150	100	100	100	75
(b) 12.5 mm cement-sand plaster		100	100	100	100	150	100	75	75	75
(c) 12.5 mm gypsum-sand plaster		100	100	100	100	150	100	75	75	75
(d) 12.5 mm vermiculite-gypsum plaster		100	100	100	100	100	75	75	62	62
8. Hollow concrete blocks, one cell in wall, of Class 2 aggregate:										
(a) unplastered						150	150	125	125	125
(b) 12.5 mm cement-sand plaster						150	150	125	125	100
(c) 12.5 mm gypsum-sand plaster						150	150	125	125	100
(d) 12.5 mm vermiculite-gypsum plaster						125	100	100	100	75
9. Cellular clay blocks not less than 50% solid:										
(a) 12.5 mm cement-sand plaster									100	75
(b) 12.5 mm gypsum-sand plaster									100	75
(c) 12.5 mm vermiculite-gypsum plaster						200	100	100	100	62
10. Cavity wall with outer leaf of bricks or blocks of clay, composition, concrete or sand-lime, not less than 100 mm thick and–										
(a) inner leaf of bricks or blocks of clay, composition, concrete or sand-lime	100	100	100	100	100	75	75	75	75	75
(b) inner leaf of solid or hollow concrete bricks or blocks of Class 1 aggregate	100	100	100	100	100	75	75	75	75	75
11. Cavity wall with outer leaf of cellular clay blocks as 9 above and inner leaf of autoclaved aerated concrete blocks, density 475–1200 kg/m³	150	100	100	100	100	75	75	75	75	75

Part VIII: Concrete floors

Construction and materials	Minimum thickness of solid substance including screed (in mm)	Ceiling finish for a fire resistance of–				
		4 hours	2 hours	1½ hours	1 hour	½ hour
Solid flat slab or filler joist floor. Units of channel or T section	90	25 mm V or 25 mm A	10 mm V or 12.5 mm A	10 mm V or 12.5 mm A	7 mm V or 7 mm A	nil
	100	19 mm V or 19 mm A	7 mm V	7 mm V	nil	nil
	125	10 mm V or 12.5 mm A	nil	nil	nil	nil
	150	nil	nil	nil	nil	nil
Solid flat slab or filler joist floor with 25 mm wood wool slab ceiling base	90			12.5 mm G	nil	nil
	100		nil	nil	nil	nil
	125	12.5 mm G	nil	nil	nil	nil
	150	nil	nil	nil	nil	nil
Units of inverted U section with minimum thickness at crown	63					nil
	75				nil	nil
	100		nil	nil	nil	nil
	150	nil	nil	nil	nil	nil
Hollow block construction or units of box or I section	63					nil
	75				nil	nil
	90		nil	nil	nil	nil
	125	nil	nil	nil	nil	nil
Cellular steel with concrete topping	63	12.5 mm V suspended on metal lathing or 12.5 mm A (direct)	12.5 mm G suspended on metal lathing	12.5 mm G suspended on metal lathing	12.5 mm G suspended on metal lathing	nil

V = vermiculite-gypsum plaster.   A = Sprayed asbestos in accordance with BS 3590:1970.   G = gypsum plaster.

Note: Where a column relating to ceiling finish contains no entry opposite a specification, the notional period of fire resistance specified in that column is not applicable.

1 Define a compartment.
2 Why do we need compartment walls and floors in a building?
3 What is the maximum size of a compartment for Purpose Group VI. The height of the factory does not exceed 28 metres. (See page 87)
4 Obtain the fire resistance for a two storey factory if the size is 20 m × 20 m × 10 m high. (See page 87)
5 Obtain the minimum thickness for a 1 hour fire resisting compartment wall in a framed building, if the construction is:
   a) Sand lime brick and 12·5 lightweight plaster
   b) Class 2 aggregate concrete block and 12·5 lightweight plaster.
6 Define Class 1 aggregate concrete block.
7 Define a Class 2 aggregate concrete block.
8 Using a precast concrete hollow beam floor in a framed building what thickness is needed if the floor is a 1 hour fire resting compartment floor?
9 State three regulations which relate to the ironmongery on the fire resisting door and the closing of the door.

Now consider this proposed building.

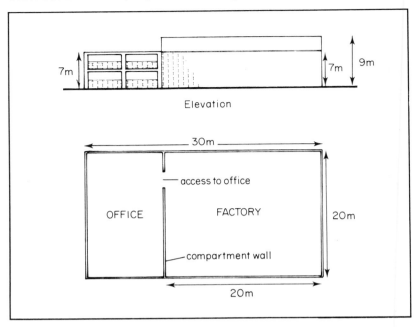

Calculate the fire resistance required for the office block and factory.

There needs to be a compartment wall separating the two purpose groups as indicated on plan. Access is required through the compartment wall between the office and the factory . . . look up Building Regulations E9 on doors. What fire resistance has the door to the factory got to have?

FIRE RESISTING DOORS

What performance requirements do you want for the fire door?

Fire doors are subject to standardised fire tests as outlined in BS 476. Remember that the fire resistance of the door was to be equal to the fire resistance of the compartment wall. There are basically two requirements of the fire door. Firstly the **integrity** of the door. By this we mean its resistance to penetration by flames or gases. Secondly its **stability**. This means it can retain its shape and size for a certain length of time.

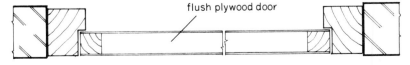

Which parts do you think are vulnerable to flame penetration?

DOOR INTEGRITY

The whole door assembly, including the frame, must be subjected to the standard BS 476 test for integrity. **Fire check** doors having resisted the passage of fire for 20 minutes are known as $\frac{1}{2}$ hour fire check doors. If under the standard test the door frame and door resist penetration of flame or gases for at least 45 minutes it is known as a 1 hour fire check door.

What is the difference between a fire check and a fire resisting door?

It depends on the time taken in a standardised test of integrity. The fire check door failed, that is it allowed flames to penetrate through the door thickness after 20 minutes, but the **fire resisting door** took at least 30 minutes for the flames/gas to penetrate, so it is classified as a $\frac{1}{2}$ fire resisting door.

DOOR STABILITY

Another performance requirement apart from integrity of door is the stability of the door assembly.

So to give you a good start in the race the door must retain its shape and size under standard tests for a certain length of time, that is its stability.

If the door does not collapse for half an hour and providing its integrity is retained for $\frac{1}{2}$ hour, the door is called a $\frac{1}{2}$ hour fire resisting door. Similarly if the whole door assembly, the frame and door retain their shape, size (ie stability) for at least 1 hour, then the door is classified as a 1 hour fire resisting door.

There are certain vulnerable parts to a fire door.

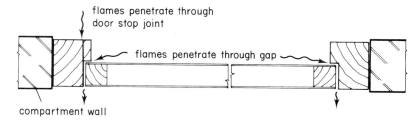

flames penetrate through door stop joint

flames penetrate through gap

compartment wall

It is recommended that the door/frame gap be a maximum of 3 mm. *Would this prevent the passage of hot gases through into the other compartment? What else is needed?*

Fire resisting door construction is outlined in British Standard 459 Part 3 *Plywood faced fire check flush doors*. The framework of the door must conform to the following sizes.

Frame member	Size mm	
	½ hr FR	1 hr FR
Stiles	38×95	38×95
Top and bottom rails	38×95	38×95
Intermediate rails	20×44	20×44
Middle rails	38×165	38×165

A skeleton framework with at least 3 mm plywood facing veneer is still not enough to give ½ hour fire protection in door integrity and stability. *What cheap sheet material can be used for good resistance to fire penetration?*

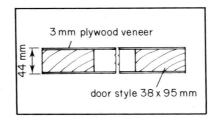

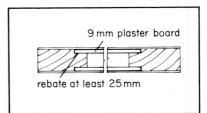

For a ½ hour fire resisting door the door thickness must be 44 mm, comprising of 2 veneer sheets of plywood and a 38 mm thick timber skeleton frame. Plasterboard or asbestos wallboard would give added protection to the flush door.

However, this construction is not sufficient to give 1 hr door integrity. The following construction is recommended for 1 hr FR doors.

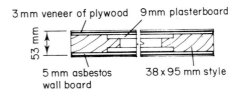

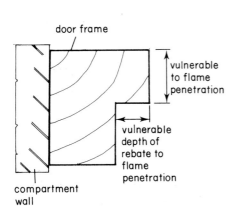

So with the addition of a 5 mm sheet of asbestos to a similar fire resistant board the door can resist the passage of fire for at least 1 hour, so the 53 mm thick door is called a 1 hr FR door.

The door will now resist fire penetration but what about the door frame and that 3 mm gap? What did you suggest to prevent hot gas passing through the door/frame gap and into the next compartment?

To ensure the door assembly is stable and keeps its integrity for the requisite fire resistance period, the door frame and rebate are to be a certain size.

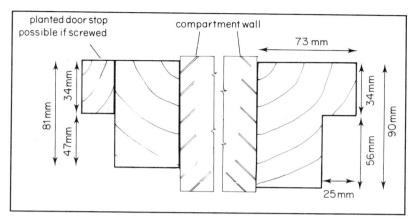

Still we have not sealed the 3 mm gap to prevent flames lipping around the door or hot gases escaping into the next compartment. An **intumescent strip** which expands under heat and forms a seal between the door and frame, is used. The intumescent strip can be either on the frame or inset in the door on a $\frac{1}{2}$ hour FR door, but on a 1 hour FR door there must be an intumescent strip on both the door frame and the door itself.

One thing we seem to have forgotten is the hanging of the door. Building Regulations stated a minimum melting point for all door furniture. What was that?

On a $\frac{1}{2}$ hour and a 1 hour fire resisting door the door is hung on $1\frac{1}{2}$ pair (3 number) of hinges and must have an automatic door closer.

Review

1   Draw the following door plan for a $\frac{1}{2}$ hr FR door; then label it and giving sizes required.

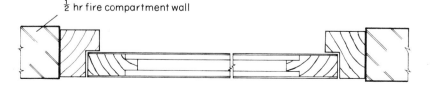

2 Draw the following diagram and indicate the size of framework for a ½ hr fire resisting door.

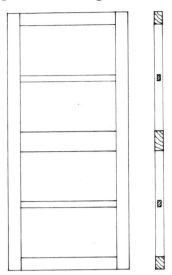

3 Explain the difference between a ½ hr fire check door and a ½ hr fire resisting door.
4 When is a compartment wall or floor required?
5 What determines the fire resistance of a fire resisting door?
6 Sketch and label with sizes and materials a 1 hour fire resisting door.
7 What kind of fire door rating is the one shown below? What is the component marked *a*?

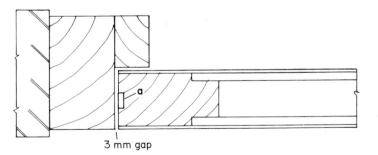

3 mm gap

18 WALL LININGS

FIRE PRECAUTIONS

You might well wonder what fire precautions have to do with ceiling and wall linings. Consider the age old problem of a burning cigarette end – the desk you are writing on, the chair you are sitting on; will they burn? A carelessly left cigarette stub could start a fire. The chair is well alight; what next? The table, curtains. The fire is well established. Flames lick the ceiling linings. Will the fire be extinguished?

Imagine the fire is below your room. Are you going to sit it out or what? You run, for it's better to be a chicken than a roasted Sunday lunch. This is one good reason for considering fire precautions when dealing with ceiling linings. Take for example a polystyrene ceiling lining. What is a good property of polystyrene tiles? What effect would this property have in a fire? Because of its thermal insulating value it *adds* to the fire's intensity. . . . We do not seem to be doing very well. If we had a noncombustible ceiling lining which did not radiate heat back into the room, are our problems solved?

Where does the flame go when it hits the ceiling? That is what we call **surface spread**. Depending on the texture of the surface, the flames could spread slowly or rapidly. This is why ceiling materials are grouped into classes to indicate their flame spread. The Building Regulations call it surface spread of fire.

THE BUILDING REGULATIONS

Read through carefully Rule E15 'Restriction of spread of flame over surfaces of wall and ceilings' and answer the questions which follow.

E15 Restriction of spread of flame over surfaces of walls and ceilings

(1) For the purposes of this regulation and the Table hereto—

   (a) CEILING includes any soffit and any rooflight or other part of a building which encloses and is exposed overhead within a room, circulation space or protected shaft;

   CIRCULATION SPACE means any space which is solely or predominantly used as a means of access between a room and a protected shaft or between either a room or a protected shaft and an exit from the building or compartment;

   ROOFLIGHT includes any domelight, lantern light, skylight or other element intended to admit daylight;

   SMALL ROOM means a room which is totally enclosed and has a floor area not exceeding that specified in column (2) of the Table to this regulation, according to the purpose group of the building or compartment; and

   TRIM means any architrave, cover mould, picture rail, skirting or similar narrow member;

   (b) any reference to the surface of a wall shall be construed as a reference to that surface including the surface of any glazing but excluding the surface of any unglazed portion of a door, any door frame, window frame, frame in which glazing is fitted, fireplace surround, mantleshelf, fitted furniture or trim;

(c) any reference to the surface of a ceiling shall be construed as a reference to that surface excluding the surface of the frame of any rooflight;

(d) any part of a ceiling which slopes at an angle of 70° or more to the horizontal and is not part of a rooflight shall be deemed to be a wall;

(e) any reference to a surface being of Class 0 shall be construed as a requirement that—

(i) the material of which the wall or ceiling is constructed shall be non-combustible throughout; or

(ii) the surface material (or, if it is bonded throughout to a substrate, the surface material in conjunction with the substrate) shall have a surface of Class 1 and, if tested in accordance with BS476: Part 6: 1968, shall have an index of performance (I) not exceeding 12 and a sub-index (i_1) not exceeding 6:

Provided that the face of any plastics material Type 1 shall not be regarded as a surface of Class 0 unless—

(a) the material is bonded throughout to a substrate which is not a plastics material and the material in conjunction with the substrate satisfies the test criteria prescribed in (ii) above; or

(b) the material satisfies the test criteria prescribed in (ii) above and is used as the lining of a wall so constructed that any surface which would be exposed if the lining were not present satisfies the said test criteria and is the face of any material other than a plastics material Type 1;

(f) any reference to a surface being of a class other than Class 0 shall be construed as a requirement that the wall or ceiling shall be so constructed that a specimen constructed to the same specification, if exposed to test by fire in accordance with BS476: Part 7: 1971, would comply with the test criteria as to surface spread of flame specified in relation to that class:

Provided that a wall or ceiling shall be deemed to have a surface of the requisite class if it is constructed to the same specification as that of a specimen which prior to 31st August 1973 was either proved to satisfy the relevant test criteria prescribed in clause 7 of BS476: Part I: 1953 or was assessed by an appropriate authority as capable of satisfying those criteria;
and

(g) in relation to a requirement that a surface shall be of a class not lower than a specified class, Class 0 shall be regarded as the highest class followed in descending order by Class 1, Class 2, Class 3 and Class 4.

(2) The surface of a wall or ceiling in a room, circulation space or protected shaft shall be of a class not lower than that specified as relevant in the Table to this regulation:

Provided that—

(a) a wall of a room may have a surface of any class not lower than Class 3 to the extent permitted by paragraph (3);

(b) an external wall of a room may have openings glazed in the manner permitted by regulation E16(2) and openings so glazed may be disregarded for the purposes of paragraph (3); and

(c) a ceiling may either have a surface of any class not lower than Class 3 to the extent permitted by paragraph (4) or may consist of plastics material to the extent permitted by regulation E16(3).

(3) Any part of the surface of a wall in a room may be of any class not lower than Class 3 if the area of that part (or, if there are two or more such parts in a room, the aggregate area of those parts) does not exceed the lesser of the following—

(a) half the floor area of the room; or

(b) (in the case of a building or compartment of purpose group I, II, or III) 20 m² or (in any other case) 60 m².

(4) Any part of the surface of a ceiling may be of any class not lower than Class 3 if that part of the surface is the face of a layer of material the other face of which is exposed to the external air and—

(a) (i) the ceiling is that of a room in a building or compartment of purpose group I, II, III, IV, V or VII or that of a circulation space in a building or compartment of any purpose group;

(ii) the area of that part does not exceed 5 m²; and

(iii) the distance between that part and any other such part is not less than 2.8 m if each part is a rooflight which complies with the provisions of paragraph (5) or 3.5 m in any other case; or

(b) (i) the ceiling is that of a room in a building or compartment of purpose group VI or VIII;

(ii) the area of that part does not exceed 5 m²;

(iii) the distance between that part and any other such part is not less than 1.8 m; and

(iv) that part and all other such parts are evenly distributed over the whole area of the ceiling and together have an area which does not exceed 20% of the floor area of the room; or

(c) the ceiling is that of a balcony, verandah, open carport, covered way or loading bay which (irrespective of its floor area) has at least one of its longer sides wholly and permanently open; or

(d) the ceiling is that of a garage, conservatory or outbuilding which (irrespective of whether it forms part of a building or is a building which is attached to another building or wholly detached) has a floor area not exceeding 40 m².

(5) The provisions referred to in paragraph (4)(a)(iii) are–

(a) that the rooflight is so designed and installed that every part of the internal surface of the light-transmitting material is above the general plane of the ceiling by no less than one quarter of the greatest dimension of that material measured internally on plan; and

(b) that any exposed internal surface (other than the frame of the rooflight) between the light-transmitting material and the general plane of the ceiling is of a class not lower than that required for the surface of the ceiling.

Table to Regulation E15

Surfaces of walls and ceilings

Purpose group of building or compartment		Maximum floor area of small room (in m²)	Class of surface for both walls and ceilings (except where separately specified)		
			Small rooms (see col.(2))	Rooms other than small rooms	Circulation spaces and protected shafts
(1)		(2)	(3)	(4)	(5)
I	Small residential: House having not more than two storeys	4	3	1 (Wall) 3 (Ceiling)	1 (Wall) 3 (Ceiling)
	Any other house	4	3	1	0
II	Institutional	4	1	0 (Wall) 1 (Ceiling)	0
III	Other residential	4	3	1	0
IV	Office	30	3	1	0
V	Shop	30	3	1	0
VI	Factory	30	3	1	0
VII	Assembly	30	3	1	0
VIII	Storage and general	30	3	1	0

1 Define circulation space.
2 Define a small room.
3 Is a room in a purpose group I, 2 storey building, having a floor plan of 1·5 m×3 m, a small room?
4 How is a class O surface defined?

5 Standardised fire tests for surface spread of fire are in accordance with which British Standard?
6 For the purpose group IV note the classes of surface spread of fire for
a) Room 10 m×4 m   *b*) Circulation space.
7 Why the difference in classes for question 6? So what is the difference between Class O and Class I surface spread of fire?

The ceiling lining materials are subject to a standardised test as dictated by BS 476. The different times taken for the flame to spread across the surface are checked and graded as follows:

Class 1 surface of very low flame spread
Class 2 surface of low flame spread
Class 3 surface of medium flame spread
Class 4 surface of rapid flame spread

These classes only measure the contribution of flame spread over a surface and *do not* adequately deal with the complex build-up of fire. Fire propagation tests are covered in BS 476 Part 6, and Class O indicates a non-combustible material. You, of course, remember the three areas where ceiling linings can help spread the fire throughout a building. Investigations are being carried out by the Fire Research Association to indicate potential fire hazards. There are three physical properties of a material which are important in a fire. Firstly, the material's thermal conductivity, secondly, the material's density and finally its thermal capacity. If we multiply these three physical properties together we get an indication of the fire hazard:

$$\frac{\text{thermal}}{\text{conductivity}} \times \text{density} \times \frac{\text{thermal}}{\text{capacity}} = \text{fire hazard potential.}$$

For example an asbestos cement ceiling under a given fire condition took 68 minutes for the surface temperature to reach 500°C. Plasterboard ceiling took only 30 minutes to reach a surface temperature of 500°C, whereas a fibre insulating board ceiling only 5 minutes. It depends on how quickly *you* want to be cooked.

Review

Before we go on to the next objective let's test your understanding of flame spread and ceiling linings.

1 State three factors affecting fire spread in ceiling linings.
2 Define according to BS 476
a) a Class 1 ceiling material
b) a Class O ceiling material
3 The flame surface spread class does not indicate fire potential. State the three factors which do indicate fire hazard potential.

DRY LINING TO WALLS

What is the benefit of dry lining to walls? The traditional method of internal finishing to masonry walls is known as a **wet trade** or finish. It takes a considerable amount of time for the plastered wall to dry before the next operation of painting can commence. So traditional plastered walls delay the house completion-time. **Dry lining** does not, but can do if we use traditional baseboard. You will remember from previous studies, the ceiling finish consisting of plasterboard and skim, using a plaster sheet called baseboard. *Why can't baseboard be used in dry lining?*

Another type of pre-finished or ready-for-decoration plaster-board is necessary. Can you name one? One of the advantages of proprietary partitions such as Paramount is its pre-finished sheet, needing no wet plaster finish. This ready-for-decoration sheet is called wallboard. British Gypsum supply wallboard in two thick-nesses, the thicknesses affecting the centres of the battening to the wall.

Why do we need battening to the wall? Can't we just fix the wall board directly to the masonry wall? What is one advantage of having a cavity between the wall and dry lining finished sheet? Some of you may have answered that the cavity acts as a barrier to rain penetration. This, of course, is correct and is one of the reasons for using dry linings in damp situations. There is, how-ever, still another reason. The cavity gives improved thermal insulation.

Size: thickness of board	Maximum centre/centre of batten
9·5 mm	450 mm
12·7 mm	600 mm

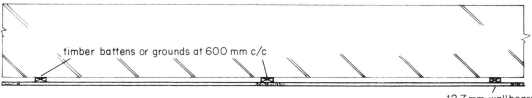

Plan through wall with dry lining

timber battens or grounds at 600 mm c/c

12·7 mm wallboard

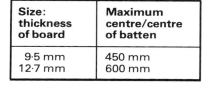

packing to make battening plumb

wallboard fixed to battening

batten plumb and aligned

brickwork out of plumb

One other reason for using battening is workmanship. Often the wall is not plumb or in alignment. Brittle wall board cannot accommodate a large variation in the background material.

Case Study

Consider a damp building. The walls are solid brick which is porous. Insulation value is minimised and decoration is constant-ly being discoloured and failing. Dry lining to all walls is recom-mended. The advantages given are reduction of dampness and increased thermal insulation. The architect recommends these materials: 9·5 mm wall board on to 25×19 mm softwood battens at 450 c/c.

What additional precaution is needed for softwood battening and why?

FIXING OF WALLBOARD

The wallboard comes in standard sheet sizes varying in width from 600 mm, 900 mm or 1200 mm. It is usually delivered in pairs with the two pre-finished faces together. How and where will you stack these wallboard sheets prior to final fixing on a housing scheme?

Taper edge boards are often used to reduce the joint finishing time.

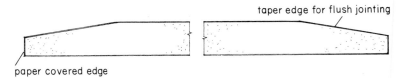

taper edge for flush jointing

paper covered edge

The boards should be fixed with the paper covered edges vertical and centred over the batten.

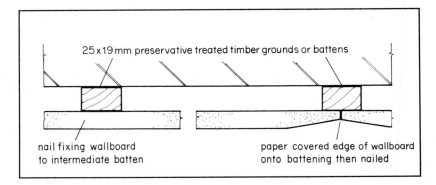

25×19 mm preservative treated timber grounds or battens

nail fixing wallboard to intermediate batten

paper covered edge of wallboard onto battening then nailed

The wallboard is now nailed to the battens at 150 mm centres, starting at the middle of the board and working out. The nails should be at least 13 mm from the edge of the board whether the edge is a paper covered edge or not. Galvanised nails are used, the size depending on the thickness of the board.

Thickness of board	Size of nail to be used
9·5 mm	30 mm long
12·7 mm	40 mm long

Exercise

Calculate the total length of battening required for dry lining a 7·00×2·40 m wall using 12·7 mm×900 mm×2400 mm wallboard.

From that exercise you will have to include a batten at the end to the cut edge of the wallboard, *but what about fixing the open edges of the board at the top and bottom of the wall?*

JOINTING AND FINISHING

Since we are using a pre-finished or ready-for-decoration wallboard we are concerned only with the joints between the wallboards. The jointing of the wallboard is exactly the same as the jointing of the Paramount partitioning. *Can you remember the sequence of jointing?* Write it down if you can.

This jointing can be done either manually or mechanically. The intermediate batten needs **nail spotting**.

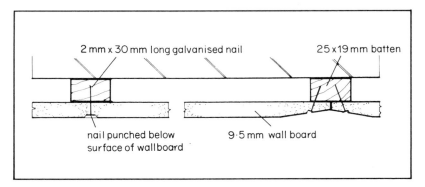

The galvanised nails are fixed at 150 mm centres and each nail needs flushing over with joint filler after it has been punched below the surface. This is called nail spotting.

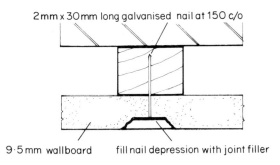

Corner junction

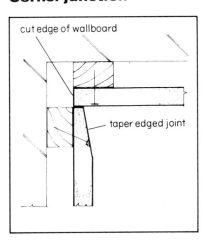

The cut board is nailed to the battening. The other corner sheet is fitted tight to the wallboard and nailed at 150 c/c using galvanised nails.

Finishing the joint

The joint is made in this sequence: apply a layer of specially prepared joint filler; next apply and press a reinforcing tape into the joint filler. This reinforcing tape is 45 mm wide. Now flush up the reinforced joint using a special joint filler. This flushing up should start immediately after the tape has been pressed home. Before the joint begins to stiffen, sponge the joint smooth. The sequence is:

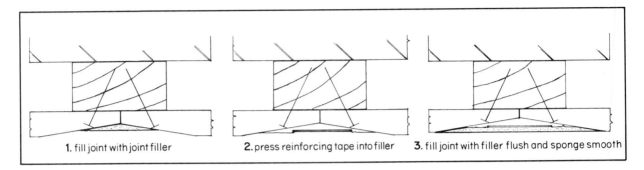

1. fill joint with joint filler   2. press reinforcing tape into filler   3. fill joint with filler flush and sponge smooth

Review

Answer these questions.

1   List the three advantages stated for dry lining walls.
2   What is the maximum battening centres for wallboard?
    a) 9·5 mm thick
    b) 12·7 mm thick
3   Softwood is often used for battening. What precautions will you take with softwood?
4   Taper edge boards are used to reduce to a minimum the jointing width. Illustrate the sequence of jointing.
5   Galvanised nails are used to fix the boarding. What size nail and centres are recommended for:
    a) 9·5 mm wallboard
    b) 12·7 mm wallboard
6   State the minimum distance the nail must be positioned from the paper covered edge of the wallboard.
7   What is nail spotting?

FIBRE BUILDING BOARD

There are basically three types of building board; hardboard, medium board, insulating board or softboard.

British Standard 1142 defines building boards as 'sheet material usually exceeding 1·5 mm thickness, manufactured from fibres of ligno-cellulosic material, with the primary bond from the felting of the fibres and their inherent adhesive properties'.

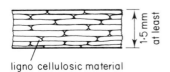

ligno cellulosic material

1·5 mm at least

What does ligno-cellulosic mean? It basically means woody. In this definition it includes both wood and other vegetable fibres.

MANUFACTURE OF BUILDING BOARD

The following timbers can be used in manufacturing building board: fir, spruce, beech, poplar, pine, birch. Other ligno-cellulosic raw materials include vegetable fibres such as cane fibre, flax and hemp. We will concentrate on the wood as a raw material.

Once the logs are felled the timber can be shredded into chips. This is the first stage – wood chips. The wood chips are screened (ie sorted into sizes) and stored. The chips are now disintegrated into their constituent fibres in a defibrator. To aid defibration the wood chips are sometimes pre-heated.

The small wood fibres are now mixed with water in the pulp chest. At this stage various chemicals can be added to improve the building board performance, such as a preservative and flame retarding salts.

The water is then drawn off and the semi-dry pulp board is passed through the forming machine. At this fourth stage the boards are made to the size and thickness specified. The pulp boards are now separated, some to be made into hardboards, others into softboards or insulating boards.

Let us take the hardboard manufacture first. The pulp boards, now approximately the right size, are taken to the hot press, where they are subjected to heat and pressure. The hardboard sheets emerging from the hot press are cured and conditioned prior to despatch.

Insulating boards are not pressed but pass directly into kilns. This final process for the insulating boards removes most of the water, which was introduced at the pulp chest. Leaving the kiln, the insulating boards are colour marked and ready for despatch.

Medium boards are manufactured in a similar way, but to a lower density.

TYPES OF HARDBOARD

There are two grades of hardboard: standard hardboard and tempered hardboard. The Fibre Building Board Development Organisation (FIDOR) recommends certain applications, depending upon whether tempered or standard hardboard is being used.

Standard hardboard is recommended for ceiling and wall linings, partitioning and flooring where mild traffic is expected.

Tempered hardboard is treated during its manufacture to improve its strength and resistance to water absorption. Tempered hardboards are recommended for hard wearing and water resisting external application.

TYPES OF MEDIUM BOARD

There are two grades of medium board, subdivided by density and finally by performance levels.

Type LM   This is a low density medium board and there are two performance grades LME ('low medium extra') and LMN ('low medium normal'). The medium board LME is the higher performance board.

Type HM   Generally the density of this type of medium board falls between 560 to 800 kg/m³ whereas the LM board density was less than 560 kg/m³. Again there are two grades of HM medium board, HME ('higher medium extra') and HMN ('higher medium normal'). The HME grade is superior. Some imported medium boards are often called 'panelboards'.

Types HME and LME are suitable for external application and are recommended for core materials to plastic laminates, wall and ceiling panelling and partitioning.

TYPES OF INSULATING BOARDS

Another name for insulating board is softboard, and softboards are made mainly from woodfibre. The density does not exceed 350 kg/m³ or its thermal conductivity 0·058 W/m°c.

There are a variety of applied finishes. The recommended applications of insulating board are thermal insulation for walls and roofs, ceiling and wall linings, and partitioning.

Review

We have now dealt with the three types of fibre building board.

Can you now answer these questions?
1   Describe the materials used for making a building board.
2   Where would standard and tempered grades of hardboard be used?
3   What types of medium board are available? Indicate which are recommended for external use.
4   What applications are insulating boards used for?

19 FACTORY ROOFS

We are concerned mainly with single storey factory and warehouse roofs. You will see numerous examples of this type of roof on industrial estates.

ROOF STRUCTURES

The roof truss shown below has to take three loadings – wind, imposed and dead loadings. The principle of triangulation of forces requires that some of the truss members will be in compression (**struts**) and some will be in tension (**ties**). The members will be mild steel angles or tees. *Can you remember the grade of mild steel used for structural work?*

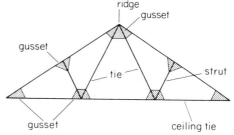

Gusset plates are used to connect the rolled steel sections. The diagrams below show an eaves gusset. Both the gusset plate and base plate are made out of flat mild steel grade 43B. The roof truss will have to be bolted to either a wall or a mild steel universal column, and accommodation needs to be made for expansion and contraction of the truss. If the load is transferred directly on to the brickwork, a **concrete padstone** is used to spread the load over a large area and avoid overloading the brickwork.

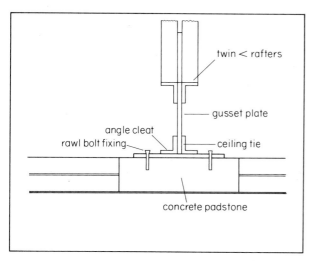

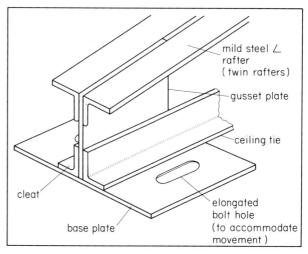

Now sketch a ceiling tie/strut gusset.

ROOF COVERING

A common type of roof covering is asbestos cement sheeting. Like most thin roof covering sheets it is profiled to give extra strength. A double skin with insulation between the sheets must be used to obtain a satisfactory 'U' value. Battens are used to stop the insulation becoming compressed.

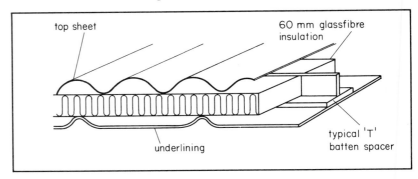

What would be the full effects of not using battens?

To achieve watertightness the sheeting must overlap at the sides and ends. A typical manufacturer's specification for side and end lapping might be as follows:

degree of exposure	roof pitch	lap treatment
Severe	17½ – 25	150mm end lap sealed
	15 – 17½	150mm end lap sealed side lap sealed
	10 – 15	300mm end lap sealed side lap sealed
Sheltered or Normal	22½ + over	150mm end lap unsealed
	15 – 22½	300mm end lap unsealed or 150mm end lap sealed
	10 – 15	150mm end lap Side and end laps sealed

What happens when the corners of four sheets meet? To avoid excessive thicknesses the sheets are mitred.

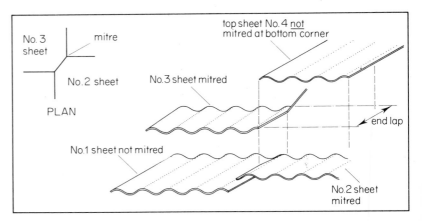

How many sheets have mitres cut at both the top left and bottom right hand corners on this plan?

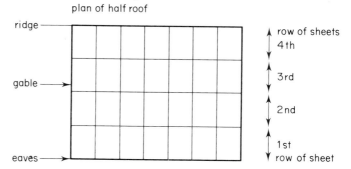

plan of half roof

ridge

gable

eaves

row of sheets
4th

3rd

2nd

1st
row of sheet

DETAILS AT EAVES, VERGES AND RIDGES

At the eaves, eaves fillers are used to close the sheeting and protect the insulation.

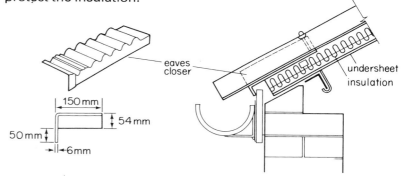

eaves closer

undersheet insulation

150 mm

54 mm

50 mm

6mm

Similarly a special accessory, called a barge board, is available to seal the roof at the verge.

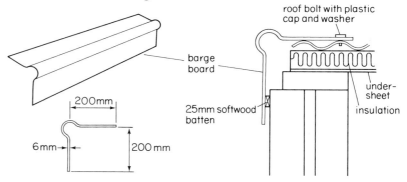

roof bolt with plastic cap and washer

barge board

under-sheet insulation

25mm softwood batten

200mm

6mm

200 mm

Finally we must seal the ridge. One way of achieving this is to use an adjustable close-fitting ridge supplied by the manufacturers. In detail it may look like this.

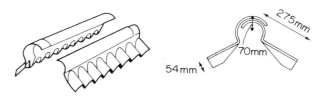

275mm

70mm

54mm

FIXING THE SHEETING

Purlins are laid across the rafters to support the sheeting.

The spacing of the purlins will depend on the roof loading, the type of sheeting used and the spacing of the mild steel roof trusses. Manufacturers will recommend maximum centres appropriate to their sheeting range.

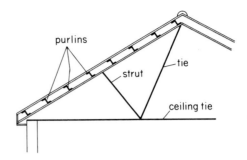

Traditionally a hook bolt was used to fix the sheeting, but this presented problems with waterproofing at the top of the bolt. An oakley clip is fixed and adjusted inside the roof and ensures a satisfactory water seal.

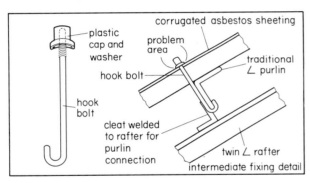

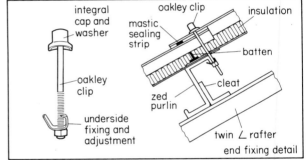

Activity

Draw these details using Zed purlins and double skin insulated asbestos cement sheeting:

1. Eaves detail.
2. Detail at strut gusset showing an end overlap of asbestos cement sheeting on a 15° severe exposure roof.
3. Explain the need for mitring and explain how this is done on asbestos cement sheeting.

20 PAINT FAILURES

This section is based on the Building Research Establishment Digest 198 (*Painting Walls, Part 2: Failures and Remedies*). Read through the Digest and answer the questions which follow.

Building Research Establishment Digest 198

Painting walls   Part 2: Failures and remedies

Principal reasons for failure (see also Table 1)

Moisture   The original construction water, water gaining access through defects in the structure or water produced by condensation.

Salts and alkali   Present in the material of construction or gaining access, for example from rising damp, by deposition as salt spray or occasionally from operations within the building.

Unsuitable surfaces   Plaster, rendering or concrete too fresh, friable or contaminated; weathered or deteriorated surfaces or paintwork.

Unsuitable conditions   Very high or low temperatures and humidities during application can seriously affect most paints. Dust and dirt produce a 'bitty' finish.

Wrong choice of paint   For a particular type of material, existing decoration, or climate. The occasions when the paint itself is not of its normal quality are quite few but some paints are less tolerant than others of adverse factors.

Moisture

The most important factor for satisfactory painting is the dryness of the wall. Excessive moisture affects the ability to apply most types of paint and affects the long-term performance. it can cause the deterioration of many materials and the movement of salts and promote the growth of moulds. Even in good conditions it may take several months to dry out a new building completely (a rough estimate being one week in drying weather for every 5 mm thickness of wet construction) and dampness often persists during the first year of occupation. Such long periods are usually unacceptable and there is a tendency to apply permanent decoration too soon. The surface of plaster or brickwork may dry quite quickly and can be decorated early provided that some restrictions on the type of decoration are acceptable, principally the inability to produce a glossy finish. Internally, on plaster, a coat of high-permeability emulsion paint is an economical temporary decoration which, unlike distemper, need not be removed later. Internally, ventilation and heat will speed up drying, and dehumidifiers can be used in very bad conditions. Ventilation is particularly important but is very often neglected, undue emphasis being placed on heating. Solid fuel appliances or other heaters vented to the outside will normally produce good ventilation but electric and central heating require extra ventilation by keeping windows open. Portable oil or gas heaters introduce additional water vapour and require maximum ventilation if they are to have any useful effect; without it they aggravate the problem and cause additional condensation on cold surfaces. Completing the painting or buildings on schedule is, of course, easier with dry forms of construction and factory-finished components.

Table 1 Painting walls—typical defects and their prevention

Defect	Cause	Undecorated		Decorated
		To prevent defect	To lessen the risk of defect	To cure after the paint has failed
Loss of adhesion; blistering and flaking of paint	(1) Water behind paint film. Defect associated with alkali attack, efflorescence or friable plaster	Allow wall to dry before decorating. Prevent rain penetration or rising damp	If wall cannot be obtained in a dry condition, use porous alkali-resistant paints	Remedy cause of dampness, strip loose paint allow to dry and redecorate
	(2) Repeated condensation causing swelling and shrinking of binder (especially found in distempers and fresh paint)	If heavy and repeated condensation cannot be avoided, use other forms of decoration	Hard gloss paints stand limited condensation well when dry. Anti-condensation paints make intermittent *light* condensation inconspicuous. Thermal insulation may help	Remedy for condensation is increased warmth with ventilation
	(3) Powdery or weak surfaces:			
	(a) *Dry-out* of gypsum plaster which loses mixing water before it has hydrated. Plaster in severe cases may be powdery or friable on the surface or throughout	Proper treatment of plaster by avoiding too rapid drying in early stages. Modern plasters rarely give trouble	If plaster is not too weak it might take a thin coat of emulsion paint alone or over a lining paper	Strip and replaster
	(b) *Delayed expansion.* If plaster suffering from dry-out later becomes wet, it may show slight rippling and softening or extensive rotting, expansion and blistering	As above	As above	As above
	(c) *Sweat-out.* If normally set gypsum plaster is wet for long periods, it loses strength	Gypsum plaster unsuitable. Conditions arise from sealing both sides of a new wet wall with impermeable finishes	As above	As above
	(d) If plasters, *other than acid Keene's,* are primed following the trowel, a thin powdery layer is formed immediately under the primer	Do not prime plaster *other than acid Keene's* until set and at least surface dry	As above	Surface is so rough it must be replastered
	(e) Poor preparation	Clean or strip powdery or dirty surfaces down to sound base	—	Strip, prepare thoroughly and repaint
	(4) Very smooth surfaces, eg glazed tiles	Degrease surfaces and use special primers. Often hard gloss paints without primers or undercoats adhere well	—	Strip and treat as new work
	(5) High suction, overall or in patches, partly due to backgrounds, trowelling and drying conditions. Lightweight plasters liable. So much of the binding medium may be removed that only, under-bound pigment is left on the surface	Correct choice of plaster and good workmanship. Use an anti-suction or alkali-resisting primer	Special primers; thin coats of emulsion paint; oil-bound distemper thinned with petrifying liquid; use of 'clearcole' or size not advised	Use 'anti-suction' primers over existing paint or after its removal
Cracking of paint	(1) Weathering—with some paints this is their ultimate normal breakdown, eg some hard gloss paints crack and flake rather than chalk or wear away	Repaint while old paint film is firmly held and continuous. Interval between repainting will depend on the paint, its situation and exposure	—	Strip cracked paint and redecorate
	(2) Repeatedly recoating surfaces	—	Emulsion paints are least prone. Where several layers of old oil-bound distemper have been built up and it is desired to avoid stripping, use a binding-down primer. Avoid hard-drying paints and enamel paints	As above

	(3) Application of hard drying gloss over soft undercoat, or application before previous coat has hardened sufficiently; application over contaminated surfaces	Ensure that finish is used with its own undercoat, or application before previous coat has hardened sufficiently; application over contaminated surfaces. Do not apply thick coats. Avoid splashes of paste or size on surfaces to be painted	—	Strip and repaint. In some cases rubbing down will remove slight crazing when the work is not affected beneath the surface
	(4) Paint of unsuitable composition even when applied to sound, dry surfaces. Associated with paint which dries hard and shrinks	—	—	Strip and repaint. Consult manufacturer
	(5) Underlying surface shrinking and cracking	Use correct plaster and mortar mixes	Use 'elastic' paints	Use elastic stopper or filler before repainting
Soft sticky films with water blisters and brown or yellow oily runs. Some pigments are bleached or discoloured	Alkali attack on oil paints and oil-bound distempers, by lime in presence of sodium and potassium salts and moisture. Cement products, lime plasters and lime-gauged plasters may cause this. Also caused by residues from alkaline paint removers	Allow to dry before painting. A few weeks subsequent maturing lessens alkalinity. One, or better two, coats of alkali-resistant primer are a useful additional precaution under oil paints. 'Alkali-killing' washes neutralise alkali only in the surface	If decorating before drying is complete, use porous alkali-resistant paints such as emulsion paints. Oil-bound distemper and size-bound distempers satisfactory in mild cases. Alkali-resistant primers lessen risk if impermeable systems (oil-based) must be used	Strip and repaint. Treat as new work
Efflorescence—patchy, crystalline or fluffy deposits sometimes outlining bricks, etc, in the background. (Not to be confused with chalking, a defect of the paint itself.)	Salts from the structure carried to the surface by water and deposited on drying. They may push off the decoration or appear over it	Allow the wall to dry out fully then remove fluffy efflorescence with a dry brush or cloth. Remove residue with a damp cloth frequently wrung out in clean water. Hard bloom needs only sanding or scraping if glossy and offering no key. Avoid water-thinned paints if possible or use them over an alkali-resistant primer if surface is heavily contaminated with salts	If decorating before drying is complete, use thin coats of permeable paints over porous alkali-resistant primers and ensure that drying conditions are good (heat and ventilation). A few emulsion paints offer a better-than-average chance of fluffy efflorescence forming on the surface of the paint without disturbing it	Strip, allow to dry and redecorate, preferably either avoiding water-thinned paints or using them over primers if the surface is heavily contaminated with salts
Coloured spots, patches or stains, often grey, black, purple, red, green or pink (on wall surfaces or paint)	Moulds which are encouraged by damp conditions and can feed on the decoration indoors or outside. Algae need damp conditions but draw little or no food from the paint; usually on exterior especially on weather side. Lichens and moss may develop in time	Remedy damp conditions. Kill organism with a fungicidal wash before painting, allow to dry and remove the organism by scrubbing or scraping	If damp conditions cannot be remedied (eg on external decorations) and there is history of mould on similar buildings, use fungicidal washes and/or fungicidal paints. Less risk with hard-drying and chemically cured paints	In mild attacks the organism may be scrubbed off, otherwise strip off old decoration. Treat the wall with a fungicidal wash. Allow to dry and redecorate with special fungicidal paints
Popping and pitting: craters blown in the surface of the plaster	Expansion of reactive particles in the plaster, usually when wet	—	—	Cut out, make good with plaster. Apply thin coat of primer
Pattern staining: patterns are outlined by variation in thickness of dust deposited on the surface	Affected mainly by surface temperature—pattern often follows variation of thermal capacity or thermal conductivity	Increase insulation to minimise temperature differences	Frequent cleaning or use of darker colours or patterns will make the defects less conspicuous	Improve insulation and repaint
Faint brown or yellow patches, soon after first decoration	Salts and stains from hollow clay tiles, some types of brick (especially if under-fired) and clinker block where a gypsum plaster* and porous paint (eg emulsion) have been used. Highly pigmented emulsion paints especially prone to this defect	Use an 'anti-suction' or alkali-resistant primer with emulsion paints, or impervious (non-emulsion) paint (preferably alkali-resistant)	Use good quality emulsion paints with a slight sheen. (Not very reliable without special primer.) Use a cement-based undercoat to the plaster	Apply (on top of existing paint) an anti-suction or alkali-resisting primer plus further emulsion paint, or flat, eggshell or gloss paint
General patchiness of sheen or colour; mortar joints visible; often occurs after further coats (usually emulsion paint)	Uneven trowelling of plaster*; variable density and suction causing varying penetration of paint binder	As above	As above	As above

*Although reported to be most frequent on lightweight plasters, this is probably only because of the predominance of these plasters in present use.

On external work drying cannot be easily accelerated and all that can be done is to protect the surface from further wetting by such means as polyethylene screens arranged to allow ventilation. Rain falling on wet or barely dry paint may cause water spotting and loss of gloss. Even when paint has been applied to a dry or apparently dry surface, moisture may later cause blistering, loss of adhesion and eventual weakening and breakdown on the film.

Measuring moisture content The condition of the surface is not always a reliable guide to the moisture content of the wall. It is sometimes possible to estimate the suitability for painting from a knowledge of the weather conditions prevailing since the wall was constructed but when in doubt it is much better to measure the moisture content directly. There are several methods available.

Weighing The most accurate method is by direct weighing of the moisture loss during oven-drying of samples obtained by drilling under controlled conditions, using appropriate corrections. A calcium carbide meter can also be used with drilled samples for on-site determinations to avoid heating. The drilling method can determine the moisture at different depths and the presence of soluble salts does not affect the results.

Hygrometer A fairly simple but reliable test is to determine the equilibrium humidity produced in an airspace in contact with the wall, using an accurate hygrometer. The space is best formed by a sealed and insulated box with a hygrometer mounted in the face opposite the wall but a simple alternative (more subject to condensation during temperature change) is to use a sheet of polyethylene fixed to the wall with adhesive tape, with the hygrometer inside. In both methods, several hours, preferably overnight, should be allowed for equilibrium to be reached.

The interpretation of the readings is given in Table 2. Coloured indicator papers which change colour according to the humidity at the surface have also been used in a similar way.

Table 2   Wall moisture content with choice of paint for early decoration

Relative humidity in equilibrium with surface %	Wall condition	Electrical meter indication (not microwave meter)	Recommendation	Suitable paint types	
				On alkaline surfaces	On neutral surfaces
100	Moisture visible	Red zone	Preferably postpone decoration and dry further; if treatment essential dry the surface before painting	Cement-based paint; possibly water-thinned epoxy paint; bituminous emulsion paint	Cement-based paint but not on gypsum plaster; possibly water-thinned epoxy paint; bituminous emulsion paint
90–100	Wet or damp patches, no obvious moisture on surface	Red zone	Preferably postpone; painting may be possible but with high risk of failure	As above; possibly emulsion-based paints	As above; emulsion paint internally
90–75	Drying, doubtful visual indication	Amber zone	Decoration possible with limitations and some risk at higher levels of moisture	Some emulsion paints; masonry paints (not chlorinated rubber); possibly epoxy paints	Most emulsion paints (except glossy); masonry paints; plaster primer; flat oil paints
Below 75	Dry	Green zone	No restriction	All oil paints on alkali-resisting primers; chlorinated rubber paint; epoxy and polyurethane paints; some emulsion paints; (alkali-resistance only as a precaution against future dampness)	Oil or emulsion paints (flat, semi-gloss and gloss); masonry paints; epoxy or polyurethane paints, one or two-pack types

Conductivity meter Two needle probes are forced into the wall and the resistance between them is measured. The resistance is reduced by soluble salts as well as by moisture, so that apparent moisture readings can be misleadingly high and a fairly dry wall may appear unfit to paint. Areas of obvious efflorescence should not be tested. If the wall is believed to be damp but readings at a shallow depth indicate low moisture content, the area should be covered with a sheet of polyethylene and the readings checked again under the sheet next day. If there is woodwork nearby it is very useful to make a check of this; as there is less likelihood of soluble salts, the reading in the wood should be more reliable than in the adjacent plaster, especially if the readings are high in the plaster and low in the wood.

Capacitance meter This has two flat electrodes which are pressed to the surface of the wall. It registers moisture only in the upper 1 or 2 mm and is inaccurate on a rough surface; soluble salts introduce errors.

Microwave meter This projects a beam of high frequency radio waves through the wall to a receiver on the other side which measures the reduction in intensity caused by the presence of moisture. The readings are for the whole thickness of the wall and are affected by the presence of soluble salts. The meter needs access to both sides of the wall, with two operators, and is expensive for infrequent measurements.

Treating moist walls A permeable coating is essential if painting has to be started while appreciable moisture is present. Water-based coatings in general are more readily applicable to slightly moist walls but must not be expected to cope with really wet conditions.

Some water-thinned epoxy based paints are claimed to be applicable to moist surfaces, especially cement, but their permeability when dry is lower than that of emulsion paint; experience of them is limited.

Attempts to hold back excessive dampness by applied coatings are not usually successful, although a number of compositions are claimed to be effective. Their use may be justified if there is an alternative route for moisture to escape to the outside and if the further access of water is unlikely or can be prevented. Metal foils can also be effective sealers if the adhesive is not affected by dampness. But applications of impervious materials to damp patches or to complete walls will usually cause the moisture to reappear elsewhere and if the opposite surface of a solid wall is also impervious, eg tiled, the coating is unlikely to adhere for long. Polyurethane sealers which are claimed to react with the moisture in the wall have been introduced but results with them can be uncertain and they have a very limited capacity for reaction.

Salts and alkali
Many types of brick contain soluble salts which it is impracticable to remove. Salts may be introduced by the use of unwashed sand, especially sea sand, in mortars and plasters, or by the addition of 'frost-proofing' additives to mortar. Rising damp or penetration by sea spray are not likely in new work but may affect redecoration.

Efflorescence is the appearance of these salts at the surface after they have been carried there in solution and the water has evaporated. The salts may be fluffy and easily removed or hard and impossible to scrape or brush off completely. On new work, efflorescence should be allowed to continue, with occasional cleaning off until no more appears. The bulky crystalline type of efflorescence is likely to disrupt impermeable paint films but the very thin, hard films of lime bloom can usually be painted over without such risk if an alkali-resisting primer or a non-saponifiable paint is used. The crystalline type often comes through emulsion paint films without much disruption but it may reduce their adhesion and more may appear with subsequent moisture movement or condensation. A useful treatment for lime bloom (on cement-based surfaces) is the application of a dilute (10 per cent) phosphoric acid solution which converts the lime to an insoluble compound. Proprietary chemicals to 'neutralise' efflorescence on plasters appear variable in action and not always reliable. Prior application of a thin penetrating coat of plaster primer often prevents further development of crystalline efflorescence. Sea salts are hygroscopic and can cause the appearance of damp patches when humidity is high; small patches may be dealt with by poultices to draw out the salts but large areas may need replastering.

Alkalinity is likely to saponify (ie soften or even liquefy) oil paints and weaken or produce white patches in some emulsion paints. All walls containing lime or cement products should be treated as being liable to cause alkali attack and either non-saponifiable paints or alkali-resistant primers underneath oil paints should be used. The masonry paints based on synthetic rubber or some acrylic solution polymers are fairly resistant and more permeable than oil-based paints used with alkali-resistant primers and can be useful in the early decoration of walls. Chlorinated rubber paints and epoxy resin paints are fully resistant to alkalies in walls but are impermeable. Some pigments are attacked by alkalies and show severe fading but paints likely to be used on cement surfaces are usually formulated with suitable pigments.

Most alkali-resisting primers are based on oils which are not easily saponified; some are based on chlorinated rubber. They vary both in their effectiveness and in their permeability to moisture. In general, the glossier the film the less the permeability; chlorinated rubber paints are least permeable per unit film thickness, and are generally semi-glossy. Because it is difficult to ensure a continuous film without misses or pin-holes, a second coat of alkali-resisting primer is advised where the risk of attack is high. Emulsion paints are usually claimed to be unaffected by alkali, at least as found in wall surfaces, but this is a variable quality and some types may be affected by cement-based surfaces and develop a lime bloom or patchy colour. Even if unaffected themselves by alkali, emulsion paints will not act as a barrier coat under susceptible paints. In general, the exterior types, usually used on cement-based surfaces, are formulated with the more alkali-resistant polymers. It is possible to improve the performance of many emulsion paints on cement rendering by a coat of alkali-resisting primer but this should be thin and penetrate the surface, rather than produce a glossy film to which emulsion paints may not adhere well.

Brown stains with no appreciable suface deposit sometimes appear on emulsion paints and are usually derived from a background of certain types of brick, hollow clay pot or clinkerblock containing soluble salts and coloured materials, or from sands containing organic matter which reacts with alkali. The best defence against this is the use of a cement-based undercoat plaster, preferably air-entrained. A coat of alkali-resisting primer is a further or alternative treatment which may also be used to correct the situation after the stains have appeared. Alternatively, flat oil paints are less susceptible to the defect.

Strong cleaning agents (washing soda, caustic soda, metal phosphates and silicates) or alkaline paint removers should not be used because they can cause saponification if not completely removed and they may be partly absorbed by porous surfaces. Neutral detergents and solvent paint removers are preferable.

Weak or unsuitable surfaces

Unsound, friable surfaces are produced by laitance on cement products or by incorrect hydration of plasters. Surfaces to be painted should be visibly sound and not powdery or crumbling. Excessively dusty and powdery surfaces provide little adhesion for paint films, particularly emulsion paints with their poor penetration, and cause a risk of peeling or flaking as well as a poor 'bitty' finish. New walls should be carefully dusted down before painting. A useful check for unsound surfaces is to apply transparent self-adhesive tape; if it shows little adhesion when pulled off and brings away loose dust and powder, the surface is unlikely to hold paint satisfactorily and should be treated. Most types of paint are affected by mould oil residues on concrete, which should be removed by solvent poultices, emulsion cleaners, neutral detergents or abrasion.

Variable or excessive porosity is often met in all types of plaster and can cause difficult or uneven application and variations in sheen, gloss or colour which persist through several coats of emulsion paint. It can often be overcome by the normal practice of thinning out the first coat with plenty of water and applying the second coat soon afterwards, but sometimes a sealer may be necessary. Alkali-resisting primers are generally suitable, but should be applied thinly so as to penetrate and not leave a glossy film. 'Plaster primers' (which are not necessarily alkali-resistant) can be used on dry, neutral walls. Oil-based paints are less affected by variable porosity, but should in any case be used over an alkali-resistant primer. Minor cracks and imperfections filled with a proprietary powder filler mixed with water may introduce areas of high suction. The difference in porosity may be evened out with a suitable sealer but it is desirable to mix the filler with diluted emulsion paint or with an oil-based undercoat.

Existing paintwork

New films shrink on drying and the force exerted may be sufficient to detach existing paint which often has poor adhesion between coats. The adhesion may be tested by cutting through the film with a razor blade in a pattern of 2 mm squares, then firmly pressing transparent self-adhesive tape to the cut area and pulling it away. Sound paint should show very little detachment; old, brittle paints may break up during the cutting. If the original paint was water-based and not very water-resistant, gentle rubbing with a moist rag or the application of a new paint to a trial area will give some indication of any risk of trouble. Existing gloss paints should be lightly flatted or wet-abraded to ensure good adhesion of subsequent paint.

Areas of flaky paint should be thoroughly cleaned off; it may be found that paint is firmly adhering to other parts of the same wall but this also should be removed if possible because the application of fresh paint often accentuates weaknesses of adhesion. A sealer or binding-down primer is often a useful safeguard but too much reliance should not be placed on its ability to penetrate multiple layers of old paints. Paint films which have cracked deeply but are not flaking should preferably be removed because of the difficulty of obliterating the outline of the cracks. Fine hair-cracking or crazing may sometimes be filled by a knifing stopper or even an undercoat. Slightly blistered or wrinkled paint which has hardened may be left in place after rubbing down to a smooth surface. Sound paint films with a chalking surface should be thoroughly washed down or preferably cleaned with fine wet abrasive paper. Old cement-based paint is often powdery and should be washed down and treated with a penetrating sealer or adhesive primer if it is to be overpainted with a non-cement based coating.

Only a few paints are incompatible with others during application (eg epoxy or chlorinated rubber paints on oil paint) but the adhesion between very different types can be poor and it is always safer to continue the use of any paint of a type which has proved satisfactory. When the nature of the existing paint is not known or a change has to be made to achieve a different decorative effect, or for water-proofing or other reasons, a costly failure may be avoided either by painting a trial patch (which may require considerable time to show any defect) or by submitting a specimen of the existing finish for examination, which the BRE Advisory Service or any consultant would do for a fee.

Mould growth Consistently damp conditions indoors encourage growth of moulds (mildew) which creates an unsatisfactory surface for re-painting. Good ventilation rather than fungicidal paints should be the first line of attack, especially at the source of humidity, with, for example, a kitchen fan or extract hood. Removal of mould is often difficult without completely removing the paint as well; it should be scrubbed off as far as possible with a detergent and the surface treated with a fungicidal wash. For a mild outbreak it may be possible to remove the growth and prevent its recurrence by curing the cause without recourse to special paints, but where there is a history of growth or reason to expect it in spite of precautions special paints should be used. A few conventional paints (emulsion or gloss) are available with added fungicides; other manufacturers produce more active

paints of different types, particularly for use in breweries, bakeries, etc. There is a risk that the latter will be incompatible not only with existing paint but also with other types that may be used in the future. If there is any likelihood of contamination of foodstuffs or if the paint can be licked by children, paint containing only non-toxic fungicides should be used. Sometimes it may be sufficient to use a paint which is relatively insensitive to water, eg a chlorinated rubber paint or an epoxy resin finish, without any fungicide present, but these should not be applied to existing paints.

Unsuitable conditions during painting
The drying of paints, whether water-thinned or oil-based, is retarded by low temperatures, high humidity and poor ventilation. It is the wall surface rather than the air temperature which is the deciding factor and it should be remembered that the wall surface may be cooler than the air in the morning and during a thaw, when condensation may also be present.

Emulsion paints, including emulsion gloss paints, may show very poor drying properties when the air is cold and damp; below 5°C they may fail to coalesce and thus not form a cohesive film; a rise in temperature after application may be unable to prevent a film failure. High humidity at low temperatures may lead to the paint running and separating on the wall. Paints based on solution polymers (eg chlorinated rubber and synthetic resins without drying oils) are rather better in cold, damp conditions. Oil-based primers, undercoats and gloss paints are usually slow to dry at low temperatures and pick up dirt and grit while they are tacky but will be sound when they eventually harden. Single-pack polyurethane gloss finishes are considerably better in cold, damp conditions than most ordinary (alkyd) gloss paints. Chemically cured finishes are slow to dry at low temperatures and must only be used according to the manufacturer's recommendations.

On porous surfaces in very hot or dry conditions, emulsion paint may be difficult to apply and may powder after drying. In these conditions, pre-wetting the wall is permissible.

Now answer these questions:
1 List the five principal reasons for paint failure.
2 Give at least three examples each of
   a) unsuitable surfaces;   *b*) unsuitable conditions.
3 What is the most important factor for satisfactory painting?

4 *a*) A portable oil/gas heater is being used to dry out a building. It is important to provide something – what?
b) Oil/gas heaters increase the water vapour in the air. What will they create where there are cold surfaces?

5 The condition of the surface is not always a reliable guide to the moisture content of the wall. List five methods of measuring directly the moisture content.

6 Using Table 2 on page 166 state the recommendation for painting the wall with the following conditions:

(relative humidity)

a) Wet/damp patches no obvious moisture.   RH 95
b) Drying, doubtful visual indication.   RH 80
c) Surface dry.   RH 60

7 Efflorescence is the appearance of salts at the surface after they have been carried there in solution and the water has evaporated. What should be done on new work if efflorescence occurs?

8 What is the bulky crystalline type of efflorescence likely to disrupt?

9 If the very thin hard films of lime bloom appear, what type of primer is used to allow painting to continue?

10 What is saponification?

11 Saponification is due to alkali surfaces. What two products are liable to cause alkali attack in walls?

12 Variable or excessive porosity is often met in all types of plaster and can cause difficult or uneven application, variations of sheen, gloss or colour. In normal practice what is this overcome by?

13 On existing paintwork how may adhesion of existing paint be tested?

14 Using the above test how would you know if the paint was sound?

15 What should you do with areas of flaky paint?

16 Paint films which have cracked deeply but are not flaking should preferably be removed. Why?

17 Consistently damp conditions indoors encourage growth of moulds (mildew). What does this create?

18 How should the mould be removed?

19 The drying of paints is retarded by three things. What are they?

20 Rather than the air temperature as the deciding factor, what temperature should be taken?

21 Using Table 1 on pages 164–5 answer the following questions. For a given 'cause' state the defect and how it can be reduced on undercoats and decorated surfaces. For example: '*Cause*, expansion of reactive particles in wet plaster; *Defect*, popping; *Undecorated preparation*, not applicable; *Decorated preparation*, cut out and make good plaster. Apply thin coat of primer.'

What then are the defects and preparations for the following?

a) *cause*: water behind paint film.
b) *cause*: application of hard gloss over soft undercoat.
c) *cause*: salts from damp structure.
d) *cause*: high suction patchiness, partly due to background.

21 CHIMNEYS

First, we will discuss the terms used then the Building Regulations applicable to fire places and flues.

FIRE SURROUND

This is basically decorative. Often it is made from tiles. It seems fashionable at the moment to make the fire surround and hearth in stone. Take away the fire surround and what is left?

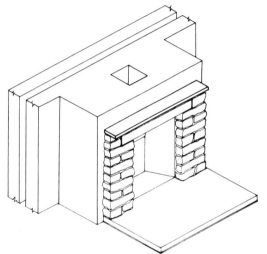

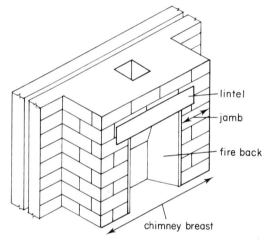

The brick construction consists of **chimney breast** and **jamb**, a precast concrete **lintel** (or throat unit) and a **fire back**.

The fire back reduces the brick fire recess and is so designed to re-direct heat back into the room. Look at the plan of the fire recess.

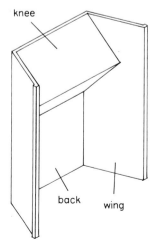

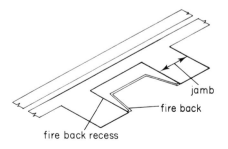

The purpose of the fire back is to contain the solid fuel, prevent the heat of the fire damaging the brick chimney, and to re-direct heat into the room.

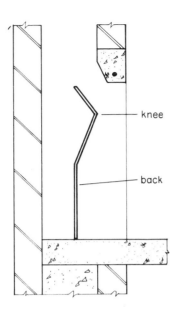

knee

back

Fire backs are made from fire clays which are clays containing a high proportion of alumina and sand. There are several terms associated with the fire back, the sides are known as **wings**, there is a **knee** and a **back**. The knee and back are best seen in this vertical section.

We must now be brave and look up the flue.

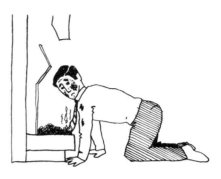

Please do not try until the fire is out!

THROAT UNIT

A vertical section through the fire place reveals a throat unit. This is usually precast concrete. The idea is to reduce the fire recess size to approximately 250×100 mm.

Why is a throat restriction needed? Much of the heat is lost up the flue. Heat loss can be reduced by this throat restriction. Additionally the velocity of the combustion gases is increased.

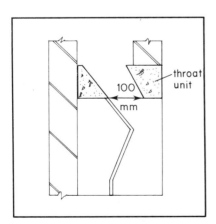

100 mm

throat unit

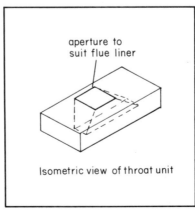

aperture to suit flue liner

Isometric view of throat unit

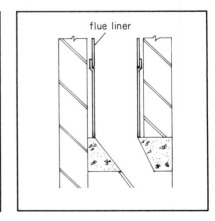

flue liner

The illustration above (centre) shows an isometric view of the throat unit. Above the throat unit is the **flue**. The size of the flue is governed by the brick size. This is called the brick flue and is normally one brick square. A flue liner is also built in as the brick flue is constructed. The flue size is governed by Building Regulations and size of fire.

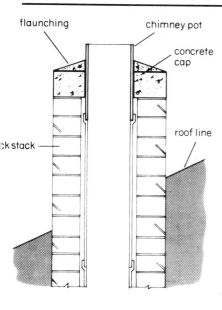

flaunching — chimney pot — concrete cap — roof line — ck stack

Now you must go outside and look at the roof because we are going to look at the stack.

Now we know the terms and components used in chimneys we can look at the Building Regulations.

Read through the 1976 Building Regulations, Part L *Chimneys, flue pipes, hearths and fireplace recess.*

PART L

Chimneys, flue pipes, hearths and fireplace recesses

L1   Application and interpretation of Part L

(1) In this Part –

APPLIANCE means –

(a) a heat-producing appliance (including a cooker) which is designed to burn –

   (i)   solid fuel (in this Part called a SOLID FUEL APPLIANCE); or

   (ii) oil (in this Part called an OIL-BURNING APPLIANCE); or

   (iii) gaseous fuel (in this Part called a GAS APPLIANCE); and

(b) an incinerator employing any means of igniting refuse, including electricity;

APPLIANCE VENTILATION DUCT means a duct forming a passage which in one part serves to convey combustion air to one or more gas appliances, in another part serves to convey the products of combustion from one or more gas appliances to the external air and intermediately serves both purposes;

CHIMNEY includes any part of the structure of a building forming any part of a flue other than a flue pipe;

CLASS I APPLIANCE means –

(a) a solid fuel appliance or oil-burning appliance having, in either case, an output rating not exceeding 45 kW; or

(b) an incinerator having a refuse combustion chamber exceeding 0·03 m³ but not exceeding 0·08 m³ in capacity,

and CLASS I shall be construed accordingly;

CLASS II APPLIANCE means –

(a) a gas appliance having an input rating not exceeding 45 kW; or

(b) an incinerator having a refuse combustion chamber not exceeding 0·03 m³ in capacity,

and CLASS II shall be construed accordingly;

CONSTRUCTIONAL HEARTH means a hearth forming part of the structure of a building;

DISCHARGE means the discharge of the products of combustion;

EXTERNAL WALL includes any external cladding or internal lining;

FLOOR includes any ceiling which is applied or fixed to the underside of the floor;

FLUE means a passage for conveying the discharge of an appliance to the external air and includes any part of the passage in an appliance ventilation duct which serves the purpose of a flue;

FLUE PIPE means a pipe forming a flue but does not include a pipe built as a lining into either a chimney or an appliance ventilation duct;

GAS FIRE means a flued gas appliance for heating one room, mainly by the emission of radiant heat, and not comprising any water heating component;

HIGH-RATING APPLIANCE means –
(a) a solid fuel appliance or oil-burning appliance having, in either case, an output rating exceeding 45 kW; or
(b) a gas appliance having an input rating exceeding 45 kW; or
(c) an incinerator having a refuse combustion chamber exceeding 0·08 m³ in capacity,
and HIGH-RATING shall be construed accordingly;

INSULATED METAL CHIMNEY means a chimney comprising a flue lining, non-combustible thermal insulation and a metal outer casing;

MAIN FLUE means a flue serving more than one appliance;

ROOF includes any ceiling which is applied or fixed to the underside of a roof and is in a plane parallel to that of the roof covering;

ROOM-SEALED APPLIANCE means a gas appliance which draws its combustion air from a point immediately adjacent to the point where it discharges its products of combustion and is so designed that the inlet, outlet and combustion chamber of the appliance, when installed, are isolated from the room or internal space in which the appliance is situated except for a door for ignition purposes;

SUBSIDIARY FLUE means a flue conveying the discharge of one appliance into a main flue; and

SUPERIMPOSED HEARTH means a hearth not forming a part of the structure of a building.

(2) (a) The provisions of this regulation and of regulation L2(1)(a), (4)(a) and (6) shall apply to the construction of a chimney which is a separate building.
(b) The provisions of this regulation and of the regulations specified in regulation L22(1) shall apply to the construction of an insulated metal chimney which serves a Class I or Class II appliance.

(c) Except as specified in this paragraph, the provisions of this Part shall not apply to chimneys described in this paragraph.

(3) Any provision in this Part which applies to a chimney, flue pipe, fireplace recess or constructional hearth serving a Class I appliance shall also apply where a solid fuel fire is intended to burn directly on a hearth without the installation of any appliance whatsoever.

(4) In relation to any Class I oil-burning appliance to which reference is made in regulation M5, regulations L3 to L7 and L10 shall not apply unless compliance therewith is required by the provisions of regulation M4.

L2   General structural requirements

(1) (a) Any chimney, flue pipe, constructional hearth or fireplace recess (whether serving a high-rating, Class I or Class II appliance) shall be –
    (i) constructed of non-combustible materials of such a nature, quality and thickness as not to be unduly affected by heat, condensate or the products of combustion; and
    (ii) so constructed and of such thickness, or, in the case of a flue pipe, so placed or shielded, as to prevent the ignition of any part of any building.
(b) Nothing in sub-paragraph (a)(i) shall prohibit –
    (i) the placing in a chimney or fireplace recess serving a Class I or Class II appliance of a damp-proof course of combustible material if it is solidly bedded in mortar; or
    (ii) the placing in a chimney or fireplace recess serving a Class I appliance of any combustible material in a position not prohibited by regulation L10; or
    (iii) the use of flue blocks having suitable combustible material incorporated during manufacture between the inner wall and surrounding material of the flue block or, if necessary to provide an expansion gap, the placing of such material between a flue lining and the surrounding material in a chimney; or
    (iv) the laying of combustible material upon the surface of a hearth in a position not prohibited by regulation L4(2).

(2) Any chimney or flue pipe (whether serving a high-rating, Class I or Class II appliance) shall be so constructed as to prevent any products of combustion escaping internally into the building.

(3) Any flue pipe (whether serving a high-rating, Class I or Class II appliance) shall –
   (a) be so placed or shielded as to ensure that, whether the pipe is inside or outside the building, there is neither undue risk of accidental damage to the flue pipe nor undue danger to persons in or about the building;
   (b) be properly supported; and
   (c) discharge either into a chimney or into the external air.

(4) (a) The outlet of any flue other than a flue described in sub-paragraph (b) shall be so situated as to prevent the discharge therefrom into the external air from entering any opening in a building in such concentration as to be prejudicial to health or a nuisance.
   (b) The outlet of a flue which serves a Class I or Class II appliance and is not the flue of a chimney which is a separate building shall comply with regulation L13 or L21 as the case may be.

(5) If provision is made for a solid fuel fire to burn directly on a hearth, secure means of anchorage for an effective fireguard shall be provided in the adjoining structure.

(6) If a flue serves an appliance which burns solid fuel or oil or is an incinerator, an opening into the flue shall be constructed so as to enable the flue to be cleaned and shall be fitted with a closely fitting cover of non-combustible material:
Provided that the requirements of this paragraph shall not apply if, while the appliance is in position, the flue is accessible for cleaning through the appliance or (if the flue communicates with a fireplace recess) through the appliance or the fireplace recess.

L3   Fireplace recesses for Class I appliances

(1) Any fireplace recess serving a Class I appliance shall have a constructional hearth which complies with the requirements of regulation L4.

(2) Subject to paragraph (3), any fireplace recess serving a Class I appliance which is constructed of bricks or blocks of concrete or burnt clay or of concrete cast *in situ* shall be so constructed that –
   (a) the jamb on each side of the recess is not less than 200 mm thick;

(b) the back of the recess is a solid wall not less than 200 mm thick or a cavity wall each leaf of which is not less than 100 mm thick; and
(c) any such thickness extends for the full height of the recess:
Provided that –
   (i) if the recess is situated in an external wall and no combustible external cladding is carried across the back of the recess, the back of the recess may be a solid wall less than 200 mm thick but not less than 100 mm thick; and
   (ii) if any part of a wall, other than a wall separating buildings or dwellings within a building, serves as the back of each of two recesses built on opposite sides of the wall, that part of the wall may be a solid wall less than 200 mm but not less than 100 mm thick.

(3) For the purposes of paragraph (2), no account shall be taken of the thickness of any part of a fireback or other appliance or the thickness of any material between an appliance and the fireplace recess.

(4) No opening shall be made in the back of a fireplace recess other than an opening which –
   (a) is made solely for the purpose of allowing the passage of convected air; and
   (b) does not communicate with a flue.

L4   Constructional hearths for Class I appliances

(1) Any constructional hearth serving a Class I appliance shall –
   (a) be not less than 125 mm thick;
   (b) (if it adjoins a floor constructed wholly or partly of combustible material, or if combustible material is laid on the hearth as a continuation of the finish of the adjoining floor in accordance with the provisions of paragraph (2)) be so constructed that any part of the exposed surface of the hearth, which is not more than 150 mm, measured horizontally, from the said floor or combustible material, is not lower than the surface of the floor and not lower than the remainder of the exposed surface of the hearth; and either
   (c) (if it is constructed in conjunction with a fireplace recess) –
       (i) extend within the recess to the back and jambs of the recess;
       (ii) project not less than 500 mm in front of the jambs; and

(iii) extend outside the recess to a distance of not less than 150 mm beyond each side of the opening between the jambs; or

(d) (if it is constructed otherwise than in conjunction with a fireplace recess) be of such dimensions as to contain a square having sides measuring not less than 840 mm.

(2) No combustible material shall be laid on a constructional hearth serving a Class I appliance as a continuation of the finish of the adjoining floor which –

(a) (if the appliance is installed directly upon or over the constructional hearth) would be nearer to the base of the appliance when installed than the distances specified in regulation M4(4); or

(b) (if the appliance is installed upon or over a superimposed hearth which complies with the requirements of regulation M4(3)(c)) would extend under the superimposed hearth to a distance of more than 25 mm or be nearer to the base of the appliance when installed than 150 mm, measured horizontally.

(3) No combustible material, other than timber fillets supporting the edges of a hearth where it adjoins a floor, shall be placed under a constructional hearth serving a Class I appliance within a distance of 250 mm, measured vertically, from the upper surface of the hearth unless such material is separated from the underside of the hearth by an air space of not less than 50 mm.

(4) Nothing in this regulation shall prohibit –

(a) the construction of a pit to hold the ash container of an appliance if –

(i) the sides and bottom of the pit are constructed of non-combustible material not less than 50 mm thick;

(ii) there is no opening in the sides or bottom of the pit other than the outlet of any duct constructed in compliance with sub-paragraph (b) or (if a side of the pit is formed by an external wall of the building) an opening situated so as to permit the removal of the container from outside the building and fitted with a closely fitting cover of non-combustible material;

(iii) no combustible material is built into a wall below or beside the pit within 225 mm of the inner surface of the pit; and

(iv) any combustible material placed elsewhere than in a wall below or beside the

pit is separated from the outer surface of the pit by an air space of not less than 50 mm; or

(b) the construction below the upper surface of a constructional hearth of a duct to be used solely for the admission of combustion air to an appliance either from outside the building or (if the floor adjoining the hearth is a floor next to the ground and is constructed as a suspended floor) from the space beneath the floor if the duct is smoke-tight and constructed of non-combustible material.

L5 Walls and partitions adjoining hearths for Class I appliances

Subject to the requirements of regulation M4(7), if any part of a wall or partition, other than a wall forming the back or a jamb of a fireplace recess which complies with the requirements of regulation L3, adjoins, or is within 150 mm of, a constructional hearth serving a Class I appliance, that part shall be constructed to a height of not less than 1·2 m above the upper surface of the hearth of solid non-combustible material not less than 75 mm thick.

L6 Chimneys for Class I appliance

(1) Any chimney serving a Class I appliance shall be either –

(a) lined with any one of the following –

(i) clay flue linings complying with BS 1181:1971; or

(ii) rebated or socketed flue linings made from kiln-burnt aggregate and high alumina cement; or

(ii) clay pipes and fittings which comply with BS 65 & 540: Part 1:1971 and are of British Standard type, socketed, imperforate and acid resistant; or

(b) constructed of concrete flue blocks made of, or having inside walls made of, kiln-burnt aggregate and high alumina cement and so made that no joints between blocks other than bedding joints adjoin any flue:

Provided that, notwithstanding the requirements of this paragraph, a chimney may be lined with a flexible flue liner if–

(i) the chimney is already lined or constructed in accordance with this paragraph; or

(ii) the chimney is not so lined or constructed but was erected under former control.

(2) Any linings or blocks described in paragraph (1) shall be jointed and pointed with cement mortar and any linings described in paragraph (1)(a) shall be so built into the chimney that the socket of each component is uppermost.

(3) If a chimney serving a Class I appliance is either –
 (a) constructed of bricks or blocks of concrete or burnt clay or of concrete cast *in situ* and in any case lined with one of the materials specified in paragraph (1)(a); or
 (b) constructed of flue blocks in compliance with paragraph (1)(b),
any flue in the chimney shall be surrounded and separated from any other flue in the chimney by solid material not less than 100 mm thick, excluding the thickness of any flue lining:
Provided that –
 (i) if the chimney forms part of a wall separating buildings or dwellings within a building and is not back-to-back with another chimney, that part of the chimney which is below the roof and separates a flue from the adjoining building or dwelling shall comprise either a solid wall not less than 200 mm thick or a cavity wall, each leaf of which is not less than 100 mm thick; and for the purposes of this sub-paragraph, any such thickness shall not include the thickness of any flue lining; or
 (ii) if the chimney forms part of an external wall and is constructed of blocks complying with paragraph (1)(b), and there is a distance of not less than 140 mm between the flue and any timber external cladding or other combustible material adjoining the outer surface of that part of the chimney which separates the flue from the external air, such part may be less than 100 mm thick but not less than 65 mm thick.

(4) If a flue in a chimney serving a Class I appliance communicates with a fireplace recess, the dimensions of every part of the flue, measured in cross-section, shall be such as will contain a circle having a diameter of not less than 175 mm:
Provided that nothing in this paragraph shall prohibit restriction of the flue to form a throat.

(5) If a flue in a chimney serving a Class I appliance does not communicate with a fireplace recess, the flue shall terminate at its lower end in a chamber which –
 (a) has means of access for inspection and cleaning fitted with a non-combustible closely fitting cover; and

 (b) is capable of containing a condensate collecting vessel.

(6) No part of a flue in a chimney serving a Class I appliance shall make an angle with the horizontal of less than 45°.

(7) Nothing in this regulation shall apply to any part of a flue in a chimney pot or other flue terminal.

L7 Flue pipes for Class I appliances

(1) No flue pipe serving a Class I appliance (whether encased or not) shall pass through any roof space, floor, internal wall or partition:
Provided that nothing in this regulation shall prohibit a flue pipe from passing through –
 (a) a floor supporting a chimney, so as to discharge vertically into the bottom of a flue in that chimney; or
 (b) a wall forming part of a chimney, so as to discharge into the side of a flue in that chimney.

(2) The cross-sectional area of any flue pipe serving a Class I appliance shall not be less than the cross-sectional area of the outlet of that appliance.

(3) For the purposes of this regulation, the expression ROOF SPACE shall not include any void between the roof covering and any ceiling which is applied or fixed to the underside of the roof and is in a plane parallel to that of the roof covering.

L8 *Deemed-to-satisfy provisions regarding materials for the construction of flue pipes for Class I appliances*

A flue pipe serving a Class I appliance shall be deemed to satisfy such requirements of regulation L2(1)(a)(i) as relate to the nature, quality and thickness of its materials if –
(a) it is constructed of cast iron complying with BS41: 1973 or of mild steel not less than 4·75 mm thick; or
(b) (being a pipe serving an appliance which is neither an open fire nor capable of being used as an open fire) any part of the pipe which is within 1·8 m of its junction with the appliance is constructed of materials specified in sub-paragraph (a) and any other part of the pipe is of heavy quality asbestos-cement complying with BS835: 1973; or
(c) (being a pipe serving a free-standing appliance which is an open fire and is not capable of

being used as a closed stove) the pipe connects the outlet of the appliance to a chimney, is not more than 460 mm long and is made of sheet steel having a thickness of not less than 1·2 mm.

L9 Deemed-to-satisfy provisions regarding placing and shielding of flue pipes for Class I appliances

(1) A flue pipe serving a Class I appliance shall be deemed to satisfy such requirements of regulation L2(1)(a)(ii) as relate to its placing or shielding if it complies with the relevant provisions of this regulation.

(2) If the flue pipe passes through a roof or external wall otherwise than for the purpose of discharging in the manner described in regulation L10(2) or (3), the flue pipe shall be –

(a) at a distance of not less than three times its external diameter from any combustible material forming part of the roof or wall; or

(b) (i) (in the case of a pipe passing through a roof) separated from any combustible material forming part of the roof by solid non-combustible material not less than 200 mm thick; or

(ii) (in the case of a pipe passing through an external wall) separated from any combustible material forming part of the wall by solid non-combustible material not less than 200 mm thick (if the combustible material is below or beside the pipe) or not less than 300 mm thick (if the combustible material is above the pipe); or

(c) enclosed in a sleeve of metal or asbestos-cement which –

(i) is carried through the roof or wall to project not less than 150 mm beyond any combustible material forming part of the roof or wall;

(ii) has between the sleeve and the pipe a space of not less than 25 mm packed with non-combustible thermal insulating material; and

(iii) (if the roof or wall is of hollow construction with an air space between the outer surface of the sleeve and any combustible material in the roof or wall) is so fitted that such material is not less than 25 mm from the outer surface of the sleeve and not less than one and a half times the external diameter of the pipe from the outer surface of the pipe; or

(iv) (if the roof or wall is of solid construction) is so fitted that any combustible material forming part of the roof or wall is not less than 190 mm from the outer surface of the pipe and is separated from the outer surface of the sleeve by solid non-combustible material not less than 115 mm thick.

(3) Where the flue pipe is adjacent to a wall or partition, it shall be at a distance of –

(a) not less than three times its external diameter from any combustible material forming part of the wall or partition; or

(b) not less than one and a half times its external diameter from any such combustible material, if such material is protected by a shield of non-combustible material which –

(i) is so placed that there is an air space of not less than 12·5 mm between the shield and the combustible material or between the shield and any non-combustible material which covers the combustible material; and

(ii) is of such width, and is fixed between the wall or partition and the pipe in such a position in relation to the pipe, that it projects on either side of it for a distance of not less than one and a half times the external diameter of the pipe.

(4) If the flue pipe passes under any floor, roof or ceiling, it shall be at a distance of –

(a) not less than four times its external diameter from any combustible material forming part of the floor, roof or ceiling; or

(b) not less than three times its external diameter from any such combustible material, if such material is protected by a shield of non-combustible material which –

(i) has an air space of not less than 12·5 mm between the shield and the combustible material or between the shield and any non-combustible material which covers the combustible material; and

(ii) is of such width and is fixed between the pipe and the floor, roof or ceiling in such a position in relation to the pipe that it projects on either side of it for a distance of not less than two and a half times the external diameter of the pipe.

L10 Proximity of combustible material – Class I appliances

(1) Subject to paragraphs (2) and (3), no combustible material shall be so placed in any chimney or fireplace recess serving a Class I appliance, or in any wall of which such a chimney or recess forms part, as to be nearer to a flue, to the inner surface of the recess, or to an opening into a flue or through the back or jambs of the recess, than 150 mm (in the case of a wooden plug) or 200 mm (in the case of any other material).

(2) Where a flue pipe serving a Class I appliance discharges into the side of a flue in a chimney, any combustible material placed in the chimney, or in any wall of which the chimney forms part, shall be separated from the flue pipe by solid non-combustible material not less than 200 mm thick (if such material is beside or below the pipe) or not less than 300 mm thick (if such material is above the pipe).

(3) Where a flue pipe serving a Class I appliance discharges into the bottom of a flue in a chimney supported by a slab, floor or roof, any combustible material forming part of or placed in the slab, floor or roof shall be separated from the flue pipe by solid non-combustible material not less than 200 mm thick.

(4) Where the thickness of solid non-combustible material surrounding a flue in a chimney serving a Class I appliance is less than 200 mm, no combustible material, other than a floorboard, skirting board, dado rail, picture rail, mantelshelf or architrave, shall be so placed as to be nearer than 38 mm to the outer surface of the chimney.

(5) No metal fastening which is in contact with combustible material shall be so placed in any chimney or fireplace recess serving a Class I appliance, or in any wall of which such a chimney or recess forms part, as to be nearer than 50 mm to a flue, to the inner surface of the recess, or to an opening into a flue or through the back or jambs of the recess.

L11 Openings into flues for Class I appliances

No opening shall be made into any flue in a chimney or flue pipe serving a Class I appliance except –

(a) an opening made for inspection or cleaning and fitted with a closely fitting cover of non-combustible material; or

(b) an air inlet which is in the same room or internal space as the appliance, is fitted with a cover of non-combustible material and is capable of being closed; or

(c) an opening which is in the same room or internal space as the appliance and is fitted with a draught stabiliser or explosion door of non-combustible material.

L12 Flues communicating with more than one room or internal space – Class I appliances

No flue in a chimney or flue pipe serving a Class I appliance shall communicate with more than one room or internal space in a building:

Provided that nothing in this regulation shall prohibit –

(a) the installation of a back-to-back grate; or

(b) the installation of two or more gas-fired incinerators in accordance with the requirements of regulation M6(2); or

(c) the making of an opening which complies with the description contained in regulation L11(a) for the purpose of giving access to a flue from a room or internal space other than that in which the appliance is installed.

L13 Outlets of flues for Class I appliances

The outlet of any flue in a chimney or flue pipe serving a Class I appliance shall be so situated that the top of such chimney or flue pipe (exclusive of any chimney pot or other flue terminal) is not less than –

(a) 1 m above the highest point of contact between the chimney or flue pipe and the roof: Provided that, where a roof has a pitch on both sides of the ridge of not less than 10° with the horizontal and the chimney or flue pipe passes through the roof at or within 600 mm of the ridge, the top of the chimney or flue pipe (exclusive of any chimney pot or other flue terminal) may be less than 1 m but not less than 600 mm above the ridge;

(b) 1 m above the top of any part of a window or skylight capable of being opened, or of any ventilator, air inlet to a ventilation system or similar opening, which is situated in any roof or external wall of a building and is not more than 2·3 m, measured horizontally, from the top of the chimney or flue pipe; and

(c) 1 m above the top of any part of a building (other than a roof, parapet wall or another chimney or flue pipe) which is not more than 2·3 m, measured horizontally, from the top of the chimney or flue pipe.

Now answer the following questions.

1 Define
   a) Class I appliance
   b) Class II appliance
   c) High rating appliance.
2 What type of material is to be used for a flue pipe, hearth or fire recess?
3 State in mm the minimum dimensions for a Class I appliance
   a) Jamb
   b) Back of recess.
4 Name three alternative materials a Class I chimney can be lined with.
5 What is the minimum size of flue for a Class I appliance?
6 State the requirements for the outlet of a flue serving a Class I appliance.
   a) Height above roof (in metres)
   b) Chimney stack within 600 mm of ridge (in mm)
   c) Above a window within 2·3 metres of chimney stack (in mm).

22 DOMESTIC HEATING

Most people find it more comfortable to have rooms heated to certain temperatures. Traditionally this was accomplished by the use of open fires in each room. This type of heating is called local heating. *Name two disadvantages of using open fire local heating in rooms.*

Suitable ambient temperatures for various rooms have been suggested by researchers. These recommendations include the heat input and the rate of air change to avoid stagnation.

Design Room Air Temperature

Room	Air Temperature	Air Change/hr
Living Room	21°C	$1\frac{1}{2}$
Dining Room	21°C	$1\frac{1}{2}$
Bedroom	13°C	$1-1\frac{1}{2}$
Hall	16°C	2
Bathroom	18°C	2
Kitchen	16°C	2

But local heating methods make it difficult to control heat input. What is the alternative to local heating?

Space heating is the same as central heating. The advantage of space heating is

1   the central boiler is used more efficiently;
2   room temperatures can be controlled better than local heating using open fires.

In these notes we are concerned with **small bore** heating systems. They are called small bore heating systems because the pipes used to distribute the heated media are only 12 mm to 22 mm diameter.

WHAT LAYOUT? The component parts of the small bore space heating system are

1   central boiler
2   distribution pipe work (small bore less than 22 mm diameter)
3   radiators in rooms.

The system illustrated is the one pipe system.

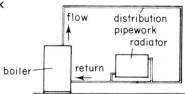

What would be the layout of a one pipe small bore space heating system for the plan below? What would the various ambient room temperatures be?

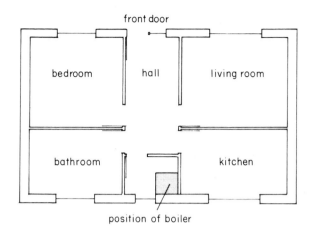

Heat loss increases with the length of pipework. The more pipework there is the greater the boiler size must be. You should aim therefore, to get minimum pipework to give maximum distribution to the strategically located radiators.

How is the water fed to the boiler? There is already a cold water storage cistern which serves the cold taps and hot water cylinder. The water authority will not allow this to be used for feeding small bore space heating systems, in order to avoid any possible contamination of drinking water. A separate cistern must therefore be used, called the **feed and expansion** cistern. Usually it has a capacity of 60 litres but this may vary with the size of the system.

The position of the cistern will affect the pressure in the system but small bore heating systems cannot rely on gravity feed supply – only a pumped circuit can cope with the small bore.

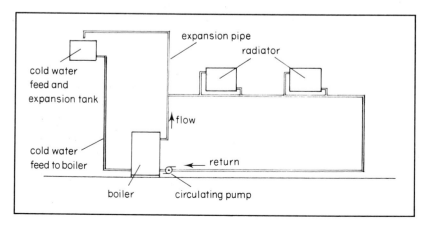

Copy this diagram and indicate on the feed and expansion tank the method of supply from the rising main and any other necessary pipes to the cistern required by the water authority.

The space heating system illustrated is an open system as opposed to a sealed or closed system.

WHAT TYPE OF BOILER?

Decisions have to be made regarding the source of fuel, and the boiler rating which will also affect the flue size required. The heating requirements are measured in kilowatts per hour. A typical small house may need a boiler rating of 20 kw/hour, whilst a 3 bedroom house may require 35 kw/hour boiler rating. These figures are only a guide.

Read the Building Regulations Part L *Chimneys, flue pipes, hearth and fireplace recesses* (pp. 173–9).

Complete the following work based on Building Regulations.

1 Define a class 1 appliance.
2 Define a high rating appliance.
3 State the thickness of a constructional hearth for a class 1 appliance.
4 State the minimum size of a square constructional hearth for a class 1 appliance.

HOW WILL THE HEATED MEDIA BE DISTRIBUTED?

Using a small bore heating system the pipe sizes are 22 mm diameter. In a one pipe system, as illustrated, both the flow and return connections of a radiator are connected to the same pipe. Some of the water will flow through the radiator, the water will lose its heat but the cooler water is returned into the flow pipe, consequently cooling the water for supplying the next radiator.

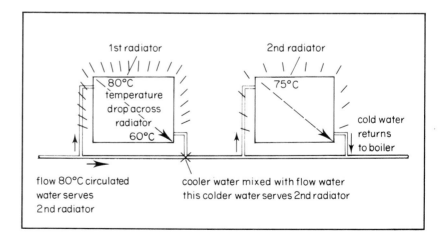

This temperature drop is for a gravity fed system. Using a pump, or forced circulation, the temperature drop across the radiator is much less, the flow temperature at the radiator will correspondingly be closer. These are obvious advantages of using a circulating pump instead of a gravity feed system.

We are looking at the one pipe small bore heating system. *Copy the diagram and fill in the labels.*

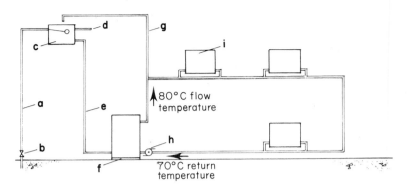

With the pump located near the boiler, the pump will push and pull the water around the system. For all practical purposes the pump 'pushes' the water at the flow outlet and pulls the water at the return. There must therefore be a point in this small bore space heating system where there is a transfer from pushing (positive pressure) to pulling (negative pressure). This is called the neutral point. Since the system is an open system there is a possibility of a leak in the system. If the leak occurs on the negative pressure of the pumped circuit (pulling water) it will cause air to be drawn into the system creating a malfunction and/or cold radiators. To avoid such low pressures and the possibility of air being drawn into the system it is necessary to ensure that the static pressure at the highest part of the circuit is more than the negative pump head. A safe rule of thumb is to have static head equal to the total pump head.

Write down the flow and return temperatures for this plan.

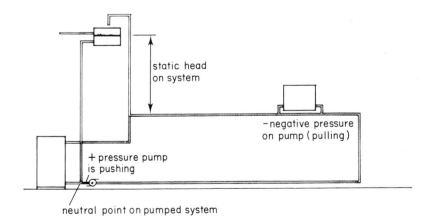

The pump must circulate the required quantity of water to maintain design temperature.

$$\frac{\text{Quantity of water}}{\text{to be circulated}} = \frac{\text{heat required}}{\text{temperature drop}}$$

The effect of bends and tees on the system must also be taken into account when calculating the pump duty (that is, the force required to keep the system at its design requirements). It is usual to add $\frac{1}{3}$ to the quantity of water to be circulated on small bore systems to allow for these bends and tee resistances.

$$\text{Pump duty} = \text{quantity of water to be circulated} + \tfrac{1}{3}$$

WHAT TYPE OF RADIATOR?

The most commonly used type of radiator is the steel panel, it is however, much less resistant to corrosion than a cast iron radiator. The panel radiator is obtainable in single, double and sometimes treble panels.

single panel radiator   double panel radiator

Makers' catalogues usually quote transmissions either as the total transmission for a given size of radiator or in kw/sq metre. For example, the room heat requirements are calculated from design temperature, heat losses through walls, ceiling, floors and heat loss due to air change are obtained. The radiator size can then be calculated.

$$\text{Radiator size} = \frac{\text{room heat requirement}}{\text{radiator emissitivity}}$$

Connections to the radiator can be either top or bottom flow.

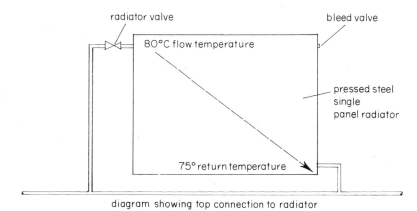

diagram showing top connection to radiator

In the top of the radiator is an air release 'bleed' valve, to release air which may collect inside. If air is trapped, and this is most likely to occur in the top radiator, the radiator will be cold.

The system is commissioned and balanced ie it is checked to make sure it is operating correctly. To balance radiators so that all radiators function correctly a lockshield valve is incorporated at each radiator. The lockshield valve is adjusted with a key during installation to balance the system, ensuring that the pump has to do the same amount of work to circulate water to each radiator, irrespective of its size, or location. It is usual to use a clip-on thermostat valve, when setting the adjustment on the lockshield valve to allow a radiator flow temperature of 80°C and return radiator temperature of 75°C.

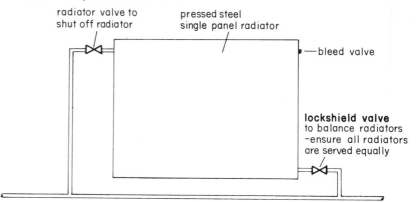

HEAT CONTROLS

The radiator valve can be used to shut off individual radiators but more precise controls are needed to benefit from space heating. One of the big advantages of a small bore heating system is its flexibility. Due to the small water content and the speed with which water can be circulated to all parts, there is an immediate response to control and heat output can be turned up or down with very little time lag. The heat output can be regulated in several ways.

1. Regulating the temperature of the water. (Adjusting the boiler thermostat.)
2. Regulating the volume of water. (Usually done with a variable flow valve which controls the amount of water flowing through.)
3. Regulating the flow of water. (Stopping and starting circulation. A room thermostat controls the pump. This is an on/off method of control.)

Can you label the components to this one pipe small bore heating system? Can you indicate the position and type of valves used on the radiators?

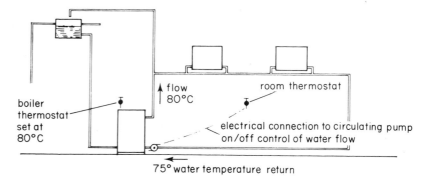

Thermostatic control of the pump is not a satisfactory method of controlling the system. No single part of the house is really representative of conditions elsewhere. Radiator thermostatic valves are best for comfort control and economy is achieved when each radiator is equipped with a thermostat.

HOW IS THE SYSTEM DRAINED?

Drain-off cocks and stop-cocks must be incorporated to allow the small bore heating system to be drained down.

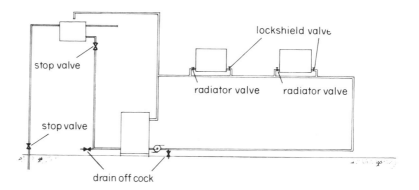

CONSTITUENT MATERIALS

Most heating installations now use pressed steel radiators and thin gauge copper tubing in combination with cast iron boilers. Defects sometimes develop after a comparatively short period of use. The problems found to be most prevalent are air locking or perforations of steel radiators.

Air locking occurs either by entrainment of air through faulty design or generation of gas within the system. Air may be drawn in through the vent pipe if this is not sufficiently high above the topmost section of distribution to balance the pump suction.

Perforation of steel radiators by pitting corrosion of the internal face is very often due to entrained air in the system.

To avoid these two problems of air locking and perforation of steel radiators, the commissioning of the system is important.

The copper tubing is a light gauge generally to BS 2871. The light gauge copper tubing can be easily bent. Special preformed fittings are also available.

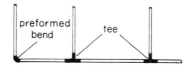

Capillary joints and non manipulative joints are most often used with light gauge copper. Pipe lengths are now manufactured to standard sizes so straight joints as well as bends and tees will have to be made.

The alternative to the capillary (or solder) joint is the non-manipulative compression joint.

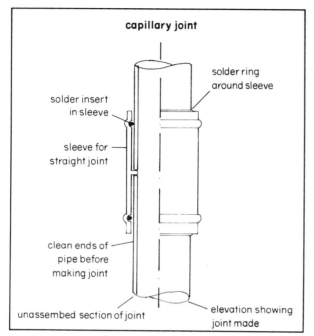

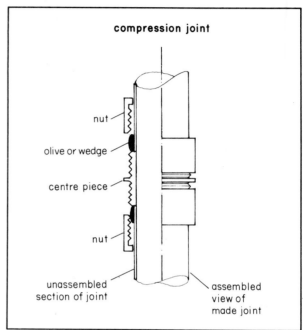

The pipework will need supporting or clipping to the wall. The recommendations from Code of Practice 310 are:

Material	Diameter	Clipping	
		Vertical	Horizontal
Light gauge copper BS 2871 table X	12 mm	1·20 m	0·800 m

One disadvantage of the one pipe small bore heating system is that the radiators return water is fed into the distribution pipework thereby cooling the water flow temperature to the next radiator.

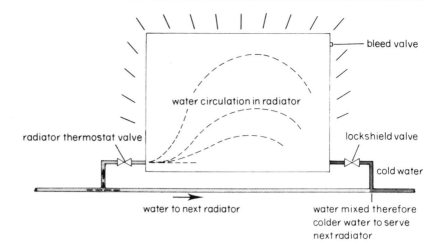

How can we improve this one pipe system? Try and work out an alternative approach. This is called the two pipe system; the return from the radiators is picked up by a separate pipe, returning to the boiler.

Review

1 Sketch and label all components of a one pipe small bore heating system, serving five radiators. Include a cold water storage tank in your diagram.
2 Sketch and label a method of controlling the temperature and flow on the two pipe system of small bore heating shown below.

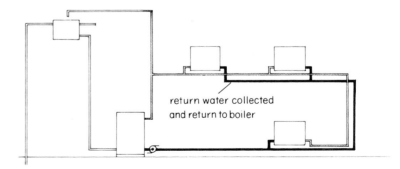

23 GAS INSTALLATION

TYPE OF GAS

Most gas used nowadays is natural gas as compared with traditional town gas. Natural gas has several advantages over town gas.

a) Higher calorific value increases the thermal capacity of the gas distribution system.

b) As no carbon monoxide is present in natural gas there is no danger of poisoning.

c) The absence of any sulphur content will reduce the possibility of acidic corrosion of natural gas appliances.

STATUTORY REGULATIONS

Gas installations have to comply with two legislating bodies; The Building Regulations and the Gas byelaws (The Gas Act 1972).

Read the Building Regulations Part L *Chimneys, flue pipes, hearths and fireplace recesses* (pp. 173–9). Now answer the following questions.

1 Define a Class II appliance.
2 What class of appliance is a gas boiler with a rating of 50 kw/hr?

Read through rules L14 to L21 of Part L of the Building Regulations opposite and answer the following questions.

3 The flue may be lined with:
a) Clay flue lining according to which BS? What is it jointed in?
b) Clay pipes according to which BS? What is it jointed in?
4 What alternative material may be used instead of a traditional brick flue with clay flue liners?
5 What is the maximum length (in metres) of circular flue serving a gas fire?
a) Flue internally situated.
b) Flue with one or more external walls.
6 What is the minimum cross sectional area (in mm²) of a flue serving a Class II gas fire?
7 What should the outlet of the flue serving a Class II appliance be fitted with?
8 What two functions is the flue terminal designed to fulfill?
9 From what must the outlet be positioned at least 600 mm?

The Gas Act (1972) deals with the layout of the pipework, material and joints to be used; size and content of meter cupboards; testing and commissioning the system.

L14   Chimneys for Class II appliances

(1) Subject to the provisions of paragraph (5), any chimney serving a Class II appliance, not being an appliance ventilation duct, shall be either –
   (a) lined with any one of the following –
      (i) acid-resistant tiles embedded in, and pointed with, high alumina cement mortar; or
      (ii) pipes which comply with specification (a) of regulation L16; or
      (iii) clay flue linings which comply with BS 1181: 1971 and are jointed and pointed with high alumina cement mortar; or
   (b) constructed of dense concrete blocks made of, or having inside walls made of, high alumina cement, and in either case jointed and pointed with high alumina cement mortar:
Provided that nothing in sub-paragraph (b) shall prohibit the use of bricks or of dense concrete blocks made otherwise than with high alumina cement, in either case jointed and pointed with cement mortar, for the construction of a chimney without flue linings if –
      (i) the flue serves one appliance only;

      (ii) the appliance served by the flue is of a type described in column (2) of the Table to this regulation; and
      (iii) the length of the flue is such as is permitted by the Table having regard to the particulars of the flue and the type of appliance specified therein.

(2) Any flue in a chimney serving a Class II appliance (including an appliance ventilation duct) shall be surrounded and separated from any other flue in the chimney by solid material not less than 25 mm thick:
Provided that where two or more flue pipes are encased in a duct, nothing in this paragraph shall require such flue pipes to be so separated.

(3) No fastening, other than a non-combustible support to a flue liner, shall be built into, or placed in, any chimney serving a Class II appliance (including an appliance ventilation duct) within 25 mm of any flue.

(4) Nothing in this regulation shall apply to any part of a flue in a chimney pot or other flue terminal.

Table to Regulation L14

Maximum length of certain flues

Situation of flue	Type of appliance	Maximum length of flue (in m)	
		If flue is circular or square, or is rectangular and has the major dimension not exceeding three times the minor dimension	If flue is rectangular and has the major dimension exceeding three times the minor dimension
(1)	(2)	(3)	(4)
(a) Flue formed by a chimney or flue pipe which is internally situated (that is to say, otherwise than as (b) below)	Gas fire	21	12
	Heater installed in drying cabinet or airing cupboard or instantaneous water heater	12	Not permitted
	Air heater or continuously burning water heater	6	Not permitted
(b) Flue formed by a chimney having one or more external walls; or by a flue pipe which is situated externally or within a duct having one or more external walls	Gas fire	11	6
	Heater installed in drying cabinet or airing cupboard or instantaneous water heater	6	Not permitted

(5) Notwithstanding the requirements of paragraph (1), a chimney serving a Class II appliance (not being an appliance ventilation duct) may be lined with a flexible flue liner if –

(a) the chimney is already lined or constructed in accordance with that paragraph; or

(b) the chimney is not so lined or constructed but was erected under former control.

L15   Flue pipes for Class II appliances

Any flue pipe serving a Class II appliance shall, if it is constructed of pipes of the spigot and socket type, have the socket of each component uppermost.

L16   *Deemed-to-satisfy provisions regarding materials for the construction of flue pipes for Class II appliances*

A flue pipe serving a Class II appliance shall be deemed to satisfy such requirements of regulation L2(1)(a) as relate to the nature, quality and thickness of its materials if it complies with any of the following specifications –

(a) clay pipes and fittings which comply with BS 65 & 540: Part 1: 1971, are of British Standard type, socketed, imperforate and acid resistant and are jointed and pointed with high alumina cement mortar; or

(b) cast iron spigot and socket flue pipes and fittings which comply with BS 41: 1973 and are coated on the inside with acid-resistant vitreous enamel and jointed with an acid-resistant compound; or

(c) sheet metal flue pipes and fittings which comply with BS 715: 1970 excluding the reference to epoxy resin from Table 2 of that publication; or

(d) stainless steel pipes and fittings; or

(e) asbestos-cement flue pipes and fittings which –

(i) comply with BS 835: 1973 or (except where they form a flue serving an incinerator) BS 567: 1973; and

(ii) (unless the flue serves one appliance only, and that appliance is of a type specified in column (2) of the Table to regulation L14, and the length of the flue is such as is permitted by that Table having regard to the particulars of the flue and the type of appliance specified therein), are coated on the inside with an acid-resistant compound which either is prepared from vinyl acetate polymer or has a rubber derivative base; and are jointed with an acid-resistant compound.

L17   *Deemed-to-satisfy provisions regarding placing and shielding of flue pipes for Class II appliances*

(1) A flue pipe serving a Class II appliance shall be deemed to satisfy such requirements of regulation L2(1)(a)(ii) as relate to its placing and shielding if –

(a) no part of the flue pipe is less than 50 mm from any combustible material; and

(b) where it passes through a roof, floor, ceiling, wall or partition constructed of combustible materials, the flue pipe is enclosed in a sleeve of non-combustible material and is separated from the sleeve by an air space of not less than 25 mm.

(2) A flue pipe serving a Class II appliance (being a pipe which is situated neither in the room or internal space in which the appliance is installed nor in an enclosed space to which no person has access) shall be deemed to satisfy such requirements of regulation L2(3)(a) as relate to the placing and shielding of a pipe within a building if –

(a) it is enclosed, either separately or together with one or more other flue pipes serving Class II appliances, in a casing constructed of suitable, but not necessarily imperforate, non-combustible material;

(b) there is a distance of at least 25 mm between the inside of the casing and the outside of any flue pipe; and

(c) no combustible material is built into, or enclosed within, the casing.

L18   Sizes of flues for Class II appliances

(1) The measurements in cross-section of a flue serving a Class II appliance (except where any part of that flue is in a ridge terminal) shall be such that –

(a) no dimension is less than 63 mm; and

(b) if the flue is rectangular in section and is not in an appliance ventilation duct, the major dimension is not more than –

(i) six times the minor dimension if the flue serves only one gas fire; or

(ii) five times the minor dimension if the flue serves only one appliance other than a gas fire; or

(iii) one and a half times the minor dimension if the flue is a main flue; or

(c) if the flue is rectangular in section and is in an appliance ventilation duct, the major dimension is not more than twice the minor dimension.

(2) The cross-sectional area of a flue serving one Class II gas fire shall be not less than 12 000 mm² and the area of the aperture in any local restrictor unit in the flue shall be not less than 6000 mm².

(3) The cross-sectional area of a flue serving one Class II appliance other than a gas fire shall be not less than the area of the outlet of that appliance.

(4) The cross-sectional area of a main flue serving two Class II gas appliances (other than gas fires) installed in the same room or internal space shall be not less than the larger of the following, that is to say –
 (a) the area of the larger of the outlets of the appliances; or
 (b) the area specified in the Table to this regulation, according to the total input rating of the appliances.

(5) Subject to the requirements of regulation M10(d)(iv), the nominal cross-sectional area of a main flue serving two or more Class II appliances installed in different storeys of a building shall be not less than 40 000 mm².

(6) The cross-sectional area of a flue in an appliance ventilation duct shall be such as will ensure that the requirements of regulation M10(b)(iii) are satisfied.

Table to Regulation L18

Minimum cross-sectional area of a flue serving two Class II gas appliances (other than gas fires) installed in the same room or internal space

Total input rating of appliances (in kW)		Minimum cross-sectional area of flue (in mm²)
Exceeding (1)	Not exceeding (2)	(3)
—	13	3750
13	18	5750
18	30	7000
30	35	9000
35	45	11 500

L19   Openings into flues for Class II appliances

No opening shall be made into a flue serving a Class II appliance except –
(a) an opening made for inspection or cleaning and fitted with a gas-tight cover of non-combustible material; or

(b) (if the flue serves an appliance other than a room-sealed appliance or incinerator) an opening which is in the same room or internal space as the appliance and serves as an air inlet or is fitted with a draught diverter or a draught stabiliser.

L20   Flues communicating with more than one room or internal space – Class II appliances

(1) No flue serving a Class II appliance shall communicate with more than one room or internal space in a building except –
 (a) a flue constructed to serve two or more Class II gas appliances installed in accordance with regulation M10; or
 (b) a flue constructed to serve two or more Class II incinerators installed in accordance with regulation M11:
Provided that nothing in this paragraph shall prohibit the making of an opening as described in regulation L19(a) for the purpose of giving access to a flue from any room or internal space other than that in which the appliance is installed.

(2) A main flue serving two or more Class II gas appliances installed in different storeys of a building (being neither a flue in an appliance ventilation duct nor a flue through which the passage of the products of combustion is assisted by a mechanically operated system of extraction) shall be so constructed that –
 (a) it is not formed by a chimney comprising part of an external wall or by a flue pipe encased in a duct comprising part of an external wall or situated externally;
 (b) it is without offsets;
 (c) it is not inclined at an angle greater than 10° from the vertical; and
 (d) each appliance discharges into it by way of a subsidiary flue complying with paragraph (3).

(3) A subsidiary flue serving a Class II gas appliance, being a flue which discharges into a main flue to which paragraph (2) relates, shall –
 (a) discharge into such main flue at a point not less than 1·2 m above the outlet of the appliance which it serves; and
 (b) make an angle of not less than 45° with the horizontal except where any other angle is necessary for the purpose of connecting the subsidiary flue to the appliance or to the main flue.

L21 Outlets of flues for Class II appliances

(1) The outlet of any flue serving a Class II appliance shall be –
  (a) fitted with a flue terminal designed to allow free discharge, to minimise down-draught and to prevent the entry of any matter which might restrict the flue;
  (b) so situated externally that a current of air may pass freely across it at all times; and
  (c) so situated in relation to any opening (that is to say, any part of a window or skylight capable of being opened or any ventilator, air inlet to a ventilation system or similar opening in any roof or external wall of a building) that –
    (i) (if the appliance is a gas appliance) no part of the outlet is less than 600 mm from any opening; or
    (ii) (if the appliance is an incinerator) no part of the outlet is less than 1 m above the top of any opening situated less than 2·3 m, measured horizontally, from the outlet.

(2) The outlet of a main flue serving two or more Class II gas appliances installed in different storeys of a building (being neither a flue in an appliance ventilation duct nor a flue through which the passage of the products of combustion is assisted by a mechanically operated system of extraction) and into which each appliance discharges by way of a subsidiary flue shall be so situated that –
  (a) the outlet is not less than 6 m above any appliance served by the flue; and
  (b) where the chimney or flue pipe passes through a pitched roof, the outlet is above the level of the ridge of the roof; or
  (c) where the chimney or flue pipe passes through a flat roof, the outlet is not below the highest of the following levels –
    (i) 600 mm above the roof; or
    (ii) 600 mm above any parapet which is within 1·5 m, measured horizontally, from the outlet; or
    (iii) the level of the top of any other part of the structure which is within 1·5 m, measured horizontally, from the outlet; or
    (iv) a level corresponding to the height of any part of the structure which is at a distance exceeding 1·5 m, measured horizontally, from the outlet reduced by one third of the difference between such distance and 1·5 m.

L22 Insulated metal chimneys serving Class I or Class II appliances

(1) An insulated metal chimney serving a Class I or Class II appliance shall be so constructed as to comply with the relevant requirements of regulations L2(4) and (6), L6(4) and (7), L11, L12, L13, L18(1), (2), (3) and (4), L19, L20(1) and L21 and with the provisions of paragraph (2) of this regulation:
Provided that regulation L20(1)(a) shall have effect as though there were substituted for the reference to regulation M10 a reference to regulation M10(a).

(2) The provisions to which reference is made in paragraph (1) are as follows –
  (a) the chimney shall be constructed of components complying with BS 4543: 1970;
  (b) joints between components shall not be situated within the thickness of any wall, floor, ceiling or roof;
  (c) if the chimney serves a Class I appliance, no part of the flue shall make an angle with the horizontal of less than 60° except where necessary to connect the chimney to the appliance;
  (d) no combustible material shall be so placed as to be nearer to the outer surface of the chimney than the distance (X) adopted for the purposes of the test procedure specified in Appendix C to BS 4543: 1970;
  (e) the chimney shall be readily accessible for inspection and replacement throughout its length;
  (f) if any part of the chimney is situated within a cupboard or storage space –
    (i) that part shall be enclosed by a removable casing constructed of suitable imperforate material;
    (ii) the distance between the inside of the casing and the outside of the chimney shall be not less than the distance specified in sub-paragraph (d); and
    (iii) no combustible material shall be enclosed within the casing; and
  (g) no part of the chimney shall pass through or be attached to any building or part of a building other than a building or part in the same occupation as that within which the appliance served by the chimney is situated.

TYPES OF GAS APPLIANCES

Name three different gas appliances used in domestic work.

Local heating

Where a space heating system is not installed a local heating source can be used. Radiant convector heaters are a good example. Some of these radiant heaters incorporate back boilers for hot water and background central heating.

There are two types of gas appliances available.
1 gas radiant convector fire
2 panel fixed fire, usually with balanced flue convectors.

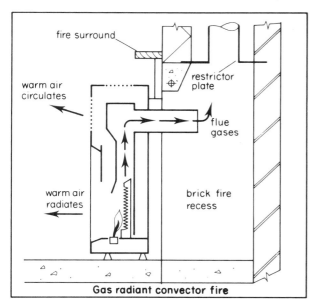

Gas radiant convector fire

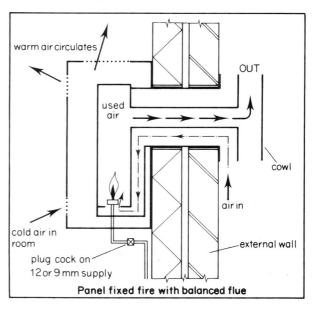

Panel fixed fire with balanced flue

The supply piping is usually 12 mm diameter.

Water heaters

Instantaneous heaters provide hot water only when it is needed. A continuously burning pilot light ignites the gas burner when the tap is turned on. Two types are available, single-point or single tap instantaneous water heaters or multi-point instantaneous water heaters. They require flues to discharge the products of combustion to the outside air.

Boilers

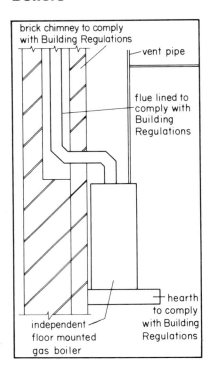

brick chimney to comply with Building Regulations

vent pipe

flue lined to comply with Building Regulations

hearth to comply with Building Regulations

independent floor mounted gas boiler

Where space heating is required a gas boiler can be fitted depending on the boiler load. The boiler may be a floor standing domestic hot water central heating boiler or a fan assisted room sealed boiler. The fan assisted room sealed boilers are very compact and can be positioned on a wall. They have balanced flues so must be placed on the outside wall.

Other domestic appliances

There are various cookers available needing a 12 mm diameter supply pipe. The gas connection to the appliance can be either fixed or flexible. The flexible connection is also often used with gas washing machines. Other domestic gas appliances include refrigerators and drying cabinets.

MATERIALS   The two materials used mainly in domestic gas installation are copper and steel.

Usually light gauge copper tube is used, made according to standards laid down by BS 2871 (Table X). The jointing is mainly capillary joints. *Sketch a section through a running joint using a capillary fitting.* The copper will need a plastic sleeve protection in certain locations.

Steel is the alternative material to light gauge copper. The steel is medium quality (colour code blue). Again the steel tube will require protection against corrosion. The protection is achieved by PVC adhesive wrapping. Running or straight joints, bends, connections, are made with threaded and screwed joints.

Before the supply is connected by the Gas Board the distribution pipework is thoroughly tested.

WHERE IS THE GAS METER?

The Gas Act Regulations determine the location of the gas meter. 'All domestic meter installations must have steel cased meters and rigid connections.' Where possible the gas meter is to be on an external wall to facilitate meter reading.

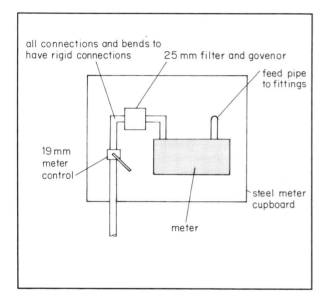

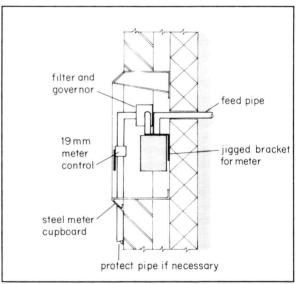

Note the rigid connections using compression couplings. *Sketch a section through a manipulative compression joint.*

The joints in the meter cupboard are manipulative compression fittings. The natural gas tends to contain a high proportion of dust and therefore filters are needed. The filter is combined with the governor.

The gas main will be at high pressure for distribution purposes. At the meter cupboard the gas pressure is reduced and a constant pressure maintained by the governor. Appliances are designed to operate within critical limits for specified gas consumption rates. Gas pressures above the specified value may result in an excess flow to the burner (called over-rating) resulting in the production of carbon monoxide due to incomplete combustion. The governor is balanced to give a domestic gas pressure of 30 millibars.

COMMISSION OF THE SYSTEM

Before connecting the supply to the meter cupboard, the gas installation is tested thoroughly by the Gas Board according to the Gas Act Safety Regulations. The Gas Board test for

a) unsafe conditions,

b) soundness test.

What constitutes unsafe conditions? There are basically three unsafe conditions: escaping gas, chimney or fire defect causing stoppage of down draught or gas spillage, and finally, incomplete combustion due to wrong adjustment of gas input or aeration.

The soundness test is as follows:

1   Meter test. Set the meter governor to register 22 millibars. Turn off meter control. Attach pressure gauge to meter outlet. Note pressure gauge reading after 2 minutes. This allows the air to settle. Open meter control until the gauge registers 22 millibars. *Do not* exceed the pressure. Wait for 1 minute. Adjust to 22 millibars pressure if necessary. Read gauge again after two minutes. The maximum pressure drop for a meter size D1 is 2·5 millibars.

2   Test the gas distribution pipework. Cap all ends and outlets except last supply. Fit to the last supply a T piece and cock. At the meter connection position and seal pressure gauge (manometer). Pump air into the system through the T piece and cock until there is a 300 millibar reading on the manometer. Allow this pressure to settle for one minute. Balance up to obtain the 300 mm water gauge. Hold test for 2 minutes. *There should be no loss of pressure.*

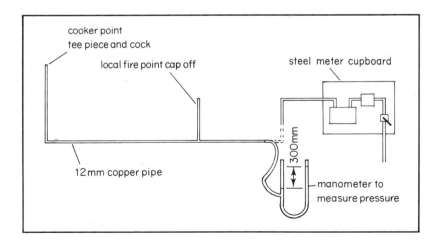

Review

Sketch and label the components to a typical domestic gas meter cupboard.

24 ELECTRICAL INSTALLATION

All electrical installation should conform to the Institute of Electrical Engineers Rules and Regulations as published in *Regulations for Electrical Equipment of Buildings*. It covers

1 the type and size of cables to be used
2 the number of appliances that can be fitted
3 the labelling of earthing and meter cupboard
4 the type of fuse capacity
5 the tests to be applied after installation.

All electrical work must conform to Regulations for Electrical Equipment of Buildings.

The number of appliances needing electricity increases the need for a sufficient number of electrical sockets. It is important to avoid the dangerous use of adaptors. Two guides regarding the number of sockets per room are available; Parker Morris Standards (usually council houses) and National House Builders Registration Council (Private Housing).

The table is a *minimum* requirement.

Electrical lighting and sockets are kept separate. The sockets for TV, kettle, etc, are known as the **power circuit**. The socket for the kettle is often called a **power point**.

Room	No. of sockets
Kitchen	3 including a cooker point
Living room	3
Dining room	3
Bedroom	2
Hall	1
Landing	1
Garage	1

LIGHTING

The Institute of Electrical Engineers Regulations limit the number of lights of one circuit to ten. What does that mean for the electrician? Suppose we have a bungalow, as is illustrated, with the number of lights in each room as shown. The total number of lights is

Living room	3
Study	1
Hall	1
Kitchen	1
Dining room	1

7 light points

The IEE allows these seven lights to be 'looped' into one circuit.

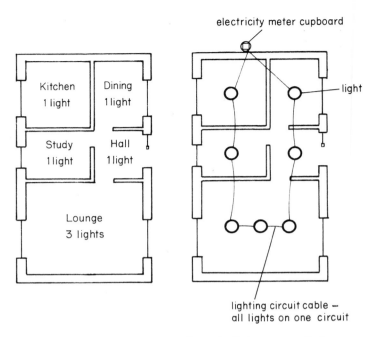

So one continuous wire is used to 'loop' all the seven lights.

What would happen if there were 12 lights required? IEE Regulations stipulate not more than 10 lights off one circuit, therefore we would need two circuits from the meter cupboard to lights.

CABLES

The cable used must be capable of carrying the required load. The Regulations for Electrical Equipment in Building has tables which give the maximum load for a given cable size. The cables are usually three wires wrapped in insulation.

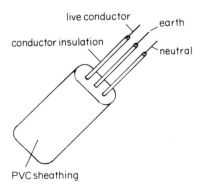

The cabling illustrated above is called twin and earth PVC cable and is suitable for lighting circuits.

Size of conductor mm²	Current rating amps
1·00	13
1·5	16
2·5	23
6·0	38

An extract of table from IEE Regulations for PVC insulated copper conductors is shown.

Before using this table we must calculate the load in amps which the cable will supply. Assume the seven lights in our bungalow were 100 watts each.

$7 \times 100 = 700$ watts capacity if all lights are switched on at the same time.

Convert watts to amps for use on table.

Current in amps $= \dfrac{power}{voltage}$ in watts (kilowatts)

$$\frac{700}{240} = 3 \text{ amp.}$$

We must now select from the available size of cable. Using the table from IEE Regulations we must use a 1·00 mm² cable.

For all lighting circuits use PVC sheathed 1·00 mm² copper conductor twin and earth.

The light point which connects the light fitting (basically the bulb and holder) to the circuit is called a ceiling rose. The switch must now be made. The switch and ceiling rose are called electrical accessories.

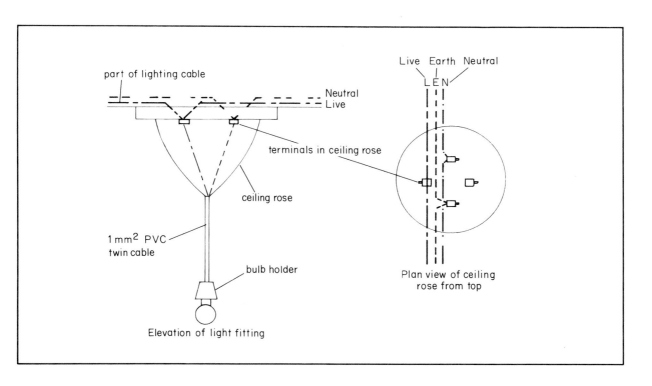

part of lighting cable

Neutral
Live

terminals in ceiling rose

ceiling rose

1 mm² PVC twin cable

bulb holder

Elevation of light fitting

Live   Earth   Neutral
L   E   N

Plan view of ceiling rose from top

The diagram below shows the wiring of the light and making of the switch. Notice that the ceiling rose has 4 terminals. The switch is made by taking from the back of the ceiling rose terminal.

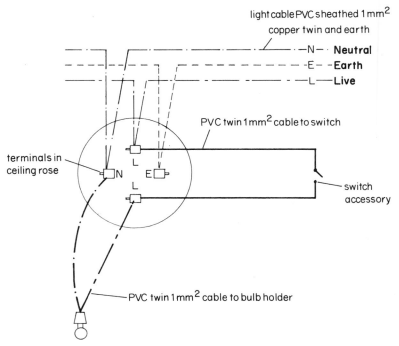

Sometimes two-way switching is required, eg for stairs and hallways for safety. This is the term given to two switches which are capable of switching the same light fitting. Pull switches must be used in bathroom to avoid the risk of electric shock from wet hands.

The lighting circuit has now been wired.

THE POWER CIRCUIT

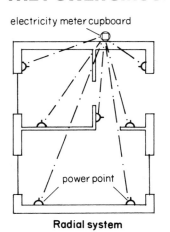

Radial system

The Regulations for Electrical Equipment of Buildings outlines the number of points off a ring main, the size of cable to be used and spurs off.

The number of power points in each room is outlined by National Housebuilding Registration Council. As with lighting, power points are in a circuit. The system shown below is a **radial system**.

The symbol ⟩ indicates a power point. Here the power points are served individually from the meter. This is expensive not only in electrician's time and materials but also because the cables are being underloaded.

Power points are all connected to one feed from the meter. This is called a **ring main**. The IEE Regulations stipulate one ring main for every 100 m² of floor area.

Fitting A in the diagram below is not in the circuit of the ring main. To include this power point might be expensive so a **spur feed** is used. The power point A is fed from the back of power point B. IEE Regulations limit the number of spurs to one per power point.

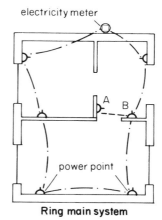

Ring main system

The usual type of power points are all **switched socket points**. They may be single or multiple (double or treble).

Some fixed appliances, like wall mounted heaters, should be connected directly to a fused socket. This is normally a spur off the ring main. The important point is that it is fused and not switched, like normal power points. The switching is done on the appliance on/off switch. Cable size is dictated by the load on the ring main.

$$\text{Load on ring main} = \frac{\text{total load of all appliances served}}{\text{potential difference ie voltage (240v)}}$$

Some electrical engineers work on the assumption that not all points will be used at the same moment in other words, that there won't be simultaneous demand. Smaller cables can then be used making installation cheaper.

The Regulations for Electrical Equipment of Buildings gives the loads that different cables can take. It is usual to use PVC flat three core cable for the ring main.

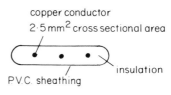

Three core flat cable

single switched
socket outlet accessory

double switched
socket outlet
accessory

IEE Regulations

Size of cable mm²	Current rating amps
1·5	16
2·5	23
6·0	38

The usual load for ring main is 20 amps excluding the cooker supply. The usual cable size is therefore 2·5 mm² conductor.

The wiring of the ring main follows a set convention. The wiring of the socket outlet (power point) will be as follows:

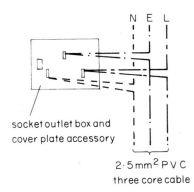

socket outlet box and
cover plate accessory

$2 \cdot 5 \, mm^2 \, PVC$
three core cable

Now that the ring main has been wired the installation is tested for correct workmanship.

Testing the wiring

The Electricity Board will test the wiring installation before connecting of the permanent supply. It should be noted that only registered electrical contractors should be used on installation work and these will test the circuit giving the houseowner or builder a test certification, besides the Electricity Board's test. Notice must be served on the Electricity Board for testing installation, as well as a request for a permanent supply (both are on specially prepared forms). The Electricity Board will test for:

1   verification of polarity
2   insulation resistance
3   test of effectiveness of earthing
4   test of ring circuit continuity.

EARTHING   **Electricity should be treated with the greatest respect.** Too many amateurs have found that out – or their relatives have.

At the generating source of electricity, the power station an electrical pressure called voltage is created; the voltage or pressure on a domestic supply is about 240 volts, adequate potential for an assassin, and far quieter than a revolver. Once the current is generated it will take the easiest way to **earth**. There are some materials which allow ready passage of the electrical current, these are called conductors, metals like copper, aluminium, and YOU. If a fault occurs in the installation, say the insulation is bared, there could easily be a fire or, if touched by a person, a serious shock.

When the socket is switched on this appliance is lethal. If you touch the fire you will 'earth', that is the current will pass through you to earth. What will that mean? A shock! Or worse still for the funeral director – someone to be laid out.

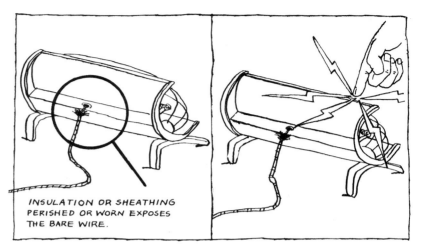

INSULATION OR SHEATHING PERISHED OR WORN EXPOSES THE BARE WIRE.

All electrical installations should be earthed at the meter cupboard. There are normally two earthing systems.

1   The house installation has an 'earth conductor', attached at the meter to a **bonded earth**. This is clearly labelled 'Do not remove earth'. This earthing is either to rods driven in the ground or to the water pipes in the house. The Electricity Board test the effectiveness of their earthing by means of an earth loop impedance test.

2   **Earth leakage** This earthing is put into the meter cupboard and is far more sensitive to shorts in the electrical circuits than the bonded earth. The earthing is normally made to the Electricity Board earthing cable and then to earth at the sub station.

FUSING

Another safety device is incorporated into the electrical system to prevent overload. If a load of 20 amps was passed through a 1 mm^2 cable the overload will be 7 amp (see Table on page 201). The insulation will be burnt (termed **bared**) and a **short** will occur with the possibility of an electrical fire. All the circuits are to be fused according to IEE Regulations.

There are normally two types of fuse now used in electrical installation:

1   cartridge
2   miniature contact breaker (mcb).

The mcb is the most sensitive device and can be easily reset after the fault is rectified. Older installations have the rewirable fuse.

The fuses are located in the meter cupboard, usually in the consumer unit.

IEE Regulations

Circuit	Fuse rating
Lighting	5 amp
Ring main	30 amp
Cooker	30 amp
Central heating to boiler	15 amp

THE METER CUPBOARD

This is now always situated on an outside wall, in a special box, to aid meter reading.

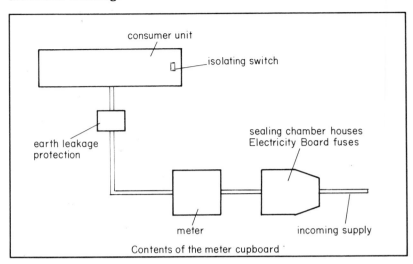

Contents of the meter cupboard

The consumer unit is often called a distribution board. The various circuits are taken from this box.

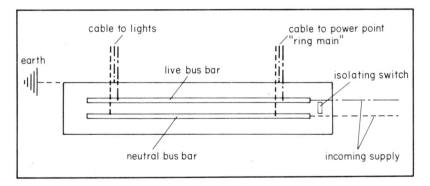

Note the box is earthed and the symbol ⑴|| is used for earth. Each circuit is taken off from this consumer unit and called a **way**. Usually there is a 6 way consumer unit installed allowing the following circuit to be served.

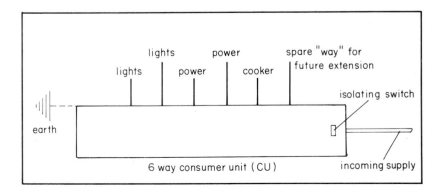

6 way consumer unit (CU)

You will remember that the fuses are housed in the consumer unit. The symbol used to indicate the fuse position is 8. The fuse must always be on the **live conductor**. After installation the Electricity Board will test to see if the fuses are on the live conductor by the test 'verification of polarity'.

Our completed view of the consumer unit looks like this.

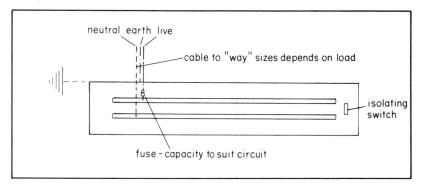

The electrical contractor will wire up the various circuits, connect the consumer unit, leaving some cable called tails for the Electricity Board to complete the meter unit.

The Electricity Board will position the earth leakage protector, meter and sealing chamber. The cable has to carry the full load required for the house so it is usual to have a PVC 10 mm² copper conductor 3 core cable, although sometimes 10 mm² aluminium conductor can be used with paper insulation.

The sealing chamber is where the joint of the house cable is connected to the Electricity Board incoming supply, which is usually 240 volt single phase supply.

The cable is laid underground so it needs to be well protected against accidental cutting. To avoid this the cable is steel wire armoured.

Housed in the sealing chamber will be the Electricity Board's fuse and cut out. The fuse is normally a 60 amp cartridge fuse.

Review

1 What should all electrical installations comply with?
2 What data is obtained from the National Housebuilding Registration Council, or the Parker Morris standards for electrical installation?
3 What are the two circuits of electricity in a house?

4 What is the maximum number of lights served on one circuit?
5 What document is used to find the right cable size to use?
6 What does the size of the cable depend on?
7 State the cable suitable for domestic lighting circuit.
8 Draw the wiring diagram for the light and switch in this position. State the cable used and the conductor size. List the electrical accessories used in this diagram.

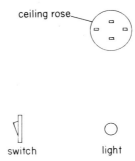

ceiling rose

switch   light

9 When is two-way switching used and for what reasons?
10 What does this symbol represent on a plan 8?
11 What is the maximum floor area for a ring main?
12 Explain when a spur is used.
13 What is the maximum number of power points served on a spur?
14 Name the cable for use in a ring main.
15 Sketch the convention for making a socket outlet.
16 Sketch the wiring diagram to suit the two switched socket outlet. State the cable to be used.
17 Name the electrical accessories used in the above diagram.
18 For what will a fused socket be used?
19 The Electricity Board must be notified of two things. What are they?
20 Explain the importance of earthing. Name the two earths on a normal electrical installation.
21 Why are circuits fused?
22 State the fuse rating for the following circuits: Lighting, Power, Cooker.
23 Name the three types of fuse available and state which of those is the most sensitive.
24 Draw a view of a 6 way consumer unit giving 2 lighting circuits, 2 ring mains, cooker and spare way. Indicate the fuse capacity required for the different ways.
   Note: The cooker cable should carry a 30 amp current. State the type of cable to be used for each circuit or way.
25 Name three pieces of property housed in the meter cupboard which belongs to the Electricity Board.
26 State the size of the incoming supply. Name the cable used for the incoming cable.
27 Before connecting the permanent supply at the sealing chamber the Electricity Board will carry out 4 tests on the installation. Name the 4 tests.
28 Draw a typical meter cupboard showing all equipment. The consumer unit is 6 way. Label the type of cabling used.

Answers

Page 32 Activity The most suitable pump for the sump pumping is the submersible pump because of its reliability. The position of the pumps and the construction programme are shown below.

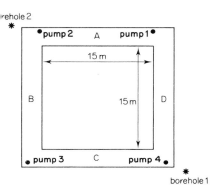

CONSTRUCTION PROGRAMME

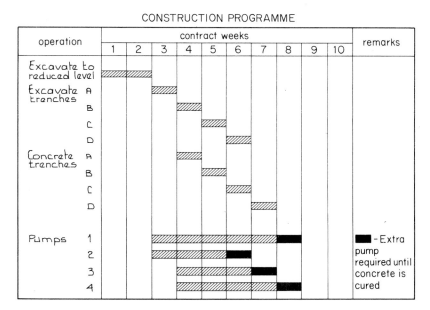

Note: For trench A pumps 1 and 2 must be working. Similarly for trench B pumps 2 and 3 must be working, etc. Pump 1 must continue working until trench D concrete curing has been completed.

Page 49 Exercise The most suitable excavator for **(a)** is the J. C. Bamford 3D or Ford 4550, and for **(b)** it would be the Poclain GY120. For **(c)** it is the Babcock and Wilcox 101MB.

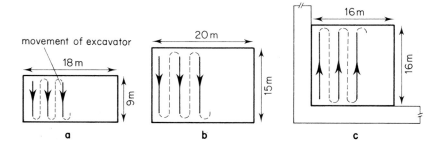

Page 54 Review 4

Number of machines needed

Section AA shows the basement depth is 5·00 m.
Quantity of excavation is therefore

$$30·50 \times 20·50 \times 5·00 = 3126·25 \text{ m}^3$$

The works programme allows three weeks for excavation.
Weekly output should therefore be

$$\frac{3126·25}{3} = 1042·08 \text{ m}^3$$

We know that one excavator's output is 14 m³ per hour.
Weekly output therefore $= 40 \times 14 = 560$ m³

$$\text{Number of machines needed} = \frac{1042}{560} = \text{'1·86'}$$

As you cannot have 1·86 machines the answer is that two machines will keep to the required programme.

Method of work and reasons

Because it is a restricted site the choice is limited to deep excavation or the dumpling method. The choice between these depends on cost comparisons.

Plant requirements

Working with two machines, the excavation should be split into two halves: Machine 1 working on gridlines 1 to 4 and A to F, machine 2 working on gridlines 4 to 7 and A to F.

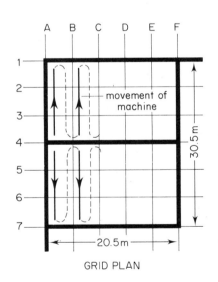

GRID PLAN

operation	contract weeks									remarks
	1	2	3	4	5	6	7	8	9	
excavation	▬	▬	▬							
r.c.wall			▬	▬	▬					
basement raft				▬	▬	▬				
erect steelwork					▬	▬	▬	▬	▬	
resources– 2 hymac 596	(2)	(2)	(2)							

Because the digging depth is 5·00 m the best choice of machine is Hymac 596.

Safety requirements

A barrier is required around the excavation. Safe access and egress should be provided for the labourers in the basement – a ladder would be sufficient. Planking and strutting should be inspected prior to use and examined regularly. The results of these inspections should be recorded in the excavation register.

Page 64 Exercise Calculate the size and main bar:

Step 1 Obtain the effective depth ratio.

$$\text{Effective depth} = \frac{\text{span}}{20} = \frac{3000}{20} = 150 \text{ mm}$$

Step 2 Assume diameter of bar reinforcement, say 2 No.12 diameter bars. Using the table 'Typical sizes of mild steel bar' on page 63, 2 No.12 diameter bars will give a cross sectional area of $2 \times 113 \cdot 1 = 226$ mm².

Step 3 Obtain overall beam size.
Assuming the cover is 25 mm to steel,

$$\text{effective depth} + \text{half bar} + \text{cover} = 150 + 6 + 25 = 181$$

Remember to take to the next step of 25 mm: beam size = 200 mm.

Therefore beam width $= \frac{2}{3} \times 200 = 133$.

The next increment is 150 mm, so overall beam size = 200 × 150 mm.

Step 4 Check assumption of 2 No.12 bars will give 0·68% area of steel to concrete.

$$200 \times 150 = \text{concrete area} = 30\,000 \text{ mm}^2$$

1% of 30 000 = 300. Now take ·68 × 300 = 204 mm². Therefore there must be at least 204 mm² of steel bar in the beam.

From step 2 we know that 2 No.12 diameter bars have 226 mm².

Answer Beam size is 200 × 150 with 2 No.12 diameter bars.

Page 82 Review **Case study 1**

1   The curing media should be located as shown below.

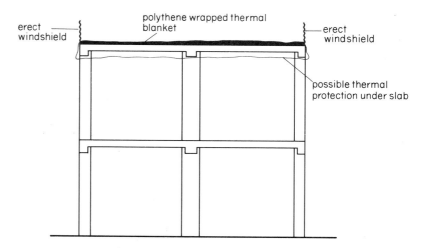

2   The diagram above shows the type of curing media. It is needed to prevent excessive heat loss in this exposed site during cold weather.

3   The curing period should be more than four days because the falling temperature will slow the curing down.

Case study 2

1   The curing media should be located as shown below.

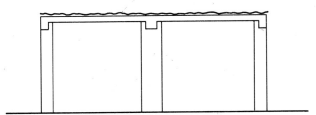

polythene or damp sacking
(alternative is to use a curing membrane)

2   The diagram above shows the type of curing media. It is needed because in dry, hot weather excessive moisture loss must be prevented.

3   The curing period should be 4 days.

4   A fine water spray should be directed at the concrete during placing to avoid plastic cracking.

Page 84 Review

1   To determine the correct type of plywood one must know how many column forms are needed. From the programme we need sufficient formwork for two days – one day for erection of formwork and one day for concrete curing. 8 columns per day are required. Therefore 16 column forms are needed.

Required number of uses from plywood face $= \dfrac{400}{16} = 25$

2   Maximum number of shutter re-uses $=$

$$\frac{\text{total columns cast}}{\text{number of forms made}} = \frac{400}{16} = 25$$

3   Chemical release agent will be used.

4   If temperature is 10°C formwork can be struck 9 hours later. This means that the formwork for each column can be struck the next day.

STAGE PROGRAMME – Shutter & concrete columns per floor

operation	contract days													remarks
	1	2	3	4	5	6	7	8	9	10	11	12	13	
erect steel reinforcement	grid line 1	2	3	4	5	6	7	8	9	10				
erect formwork		grid line 1	2	3	4	5	6	7	8	9	10			
concrete columns			grid line 1	2	3	4	5	6	7	8	9	10		
strike formwork and re-use				grid line 1	2	3	4	5	6	7	8	9	10	

Page 119 Exercise   The amount of tubular scaffolding required.

Standards
The centre of the standards is 2.1 m.

$$\frac{8\cdot4}{2\cdot1}=4+1=5 \text{ rows of 2.}$$

Therefore 10 standards are needed.

$$10\times2\cdot30 \text{ m}=23\cdot0 \text{ m.}$$

Legers
One lift of scaffold. 2 No. of 8·40 m=16·8 m.

Guardrail
One guard rail. 1 No. of 8·40 m=8·4 m.

Summary	
standards	23·0 m
ledgers	16·8 m
guardrail	8·4 m
	48·2 m

At 10p per week hire charge, this would cost £4.82p.

Page 137 Exercise   The length of battening required.

$$\text{Number of intermediate battens required}=\frac{5\cdot2}{1} \text{ (panel width)}=5$$

(not 5·2 because the cut panel will have a wall batten)

$$5\times2\cdot5=12\cdot5 \text{ m}$$

Sole batten 300 mm long: 5×·3=1·5 m

Intermediate battens	=12·5 m
Ceiling	= 5·2 m
Sole plate	= 5·2 m
Wall battens	= 5·0 m
Sole batten	= 1·5 m
Total	=29·4 m